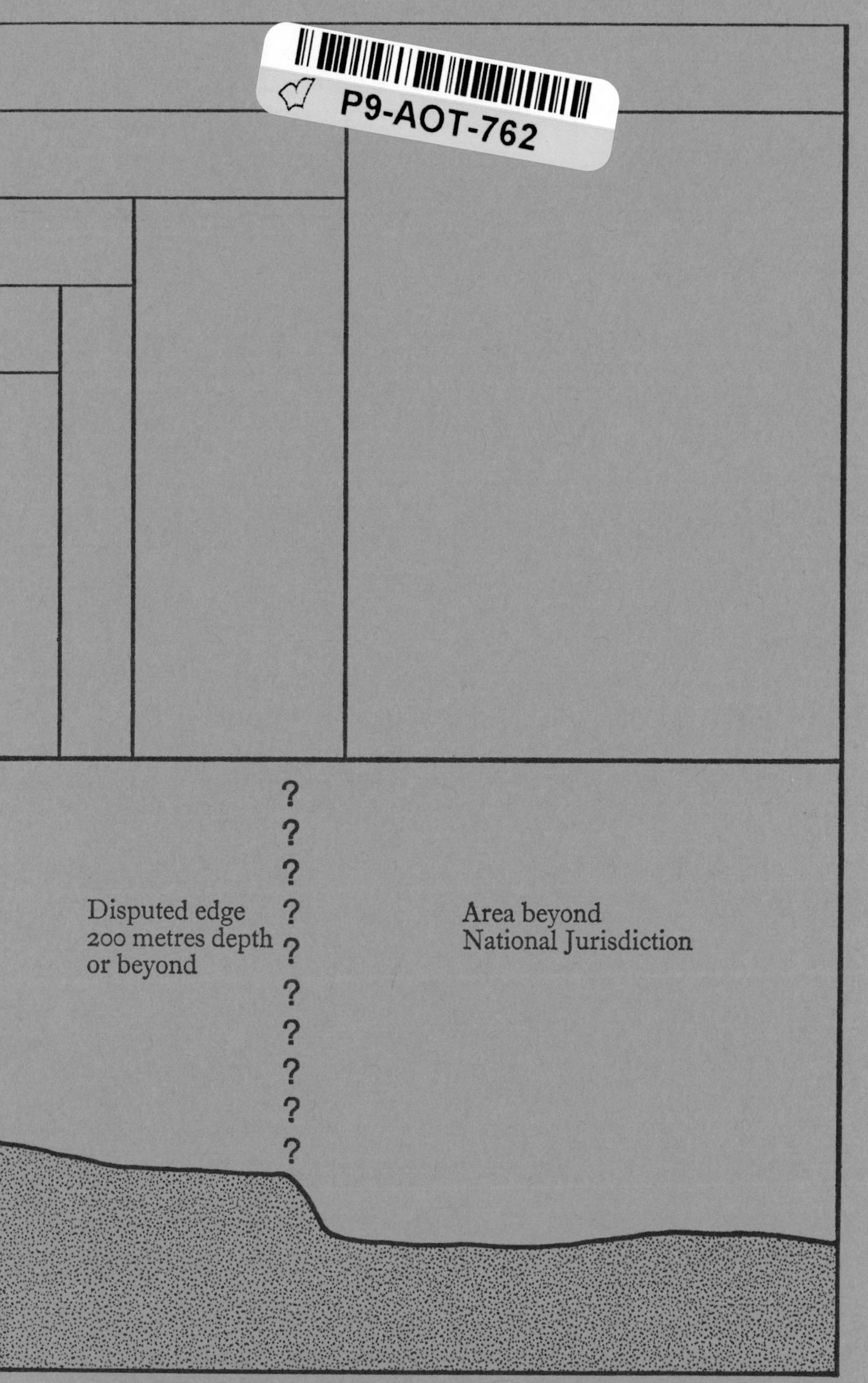

Diagram of Boundary Terms used in the Law of the Sea

The Control of the Sea-bed

ALSO BY EVAN LUARD

Britain and China (1962)
Nationality and Wealth (1964)
Conflict and Peace in the Modern International System (1968)
The Evolution of International Organizations (ed.) (1965)
The International Protection of Human Rights (ed.) (1968)
The International Regulation of Frontier Disputes (ed.) (1970)
The International Regulation of Civil Wars (ed.) (1971)
Types of International Society (1976)

The Control of the Sea-bed

AN UPDATED REPORT

EVAN LUARD

TAPLINGER PUBLISHING COMPANY
NEW YORK

First published in the United States in 1977 by
TAPLINGER PUBLISHING CO., INC.
New York, New York

Printed in Great Britain

Library of Congress Catalog Card Number: 76-11050

ISBN 0-8008-1811-3

Contents

Foreword

The international conflict now developing over the sea-bed is an issue of a kind that has never emerged before in man's history. It is arguable that it is the most important dispute that has ever arisen in dealings among states, for it concerns the ownership of two-thirds of the territory of the earth and a substantial proportion of its wealth.

It is now known that there exist on the floor of the ocean resources of enormous value: not only petroleum and natural gas, already being exploited close to the shore, but minerals—copper, nickel, cobalt, manganese and others—in unknown, but certainly enormous quantities. It has been estimated that there are many times as much of some of these materials as exist on earth itself. It is possible that they may be able to be exploited as cheaply as, or even more cheaply than, those known on land. The key question which is not yet decided is: who do they belong to?

Earlier in man's history, when new resources became known in various parts of the earth, such questions were easily, though brutally, answered. They were simply seized by those who could: by individuals, as in gold rushes or in prospecting for mines; or by nations forcibly acquiring control of vast overseas territories, as in the scramble for colonies. For the resources of the sea-bed a more civilized method of resolving the matter has so far been attempted: through discussions among governments, mainly under the auspices of the UN, concerning the way those resources should be controlled and exploited, and their revenues distributed.

There is a wide range of contentious questions to be decided. How wide is the 'continental shelf', the area in which coastal states have the undisputed right to control and benefit from exploitation? Is there, beyond that, an area that is fully international? What rights and obligations should governments enjoy in each of these areas? What weight is to be attached to unilateral claims by governments to large extents of territorial waters or continental shelf beyond the usually accepted limits? What rights do individuals or states have to undertake free exploitation in the deep-sea area beyond the shelf? Under whose authority and on what terms should operations be allowed there?

If there is to be some kind of international control of the deeper area, what form should it take? How should the revenues of exploitation be divided? Should rights of exploitation be granted to governments in the first place, or direct to companies, or will there be exploitation on an international basis? And will decisions be reached within the international system by majority vote, or on what other system of decision-making?

Again, what military uses of the sea-bed will be permitted? If there should be disarmament then how will it be enforced? How will pollution of the marine environment that may result from exploitation be prevented and controlled? How are the difficult legal questions that may arise among governments, or between governments and an international authority, to be resolved? How will decisions be imposed when governments object to them? And what kind of administration will be required to control any international area, once its limits have been agreed?

These are the types of issue that have arisen, with the rapid development of technology, over recent years. They are the issues which have been discussed among governments since 1968, and which this book seeks to consider. The book examines first the facts and the problems in four major areas: economic, legal, military and scientific. Secondly, it recounts the major discussions of the issue so far, and the various views put forward. Finally, it seeks to examine the types of solution that may finally be attainable, given the views and attitudes of the major powers and groups of powers which are involved.

The book is not mainly designed to prescribe the type of settlement that is ultimately required (though it makes some suggestions) but to *describe*, for the benefit of the general public, the nature of the issues at stake. If, as has been suggested, this is one of the most important issues in history, which will crucially determine the distribution of wealth and welfare in the world for years to come, it is important that it should be discussed far more widely, both in this country and abroad, than it has been so far.

The issue is, in the author's view, among the greatest challenges and the greatest opportunities that mankind has yet faced: a challenge, because a failure to confront it effectively and fairly will not only increase existing inequalities among nations but will also stir up huge resentments, which will divide them for long in the future; an opportunity, because a successful resolution could inaugurate a completely new era in man's history, in which resources are for the first time controlled on an international basis and their benefits widely shared. Just as nationalization has for many years been seen as a means of reducing inequalities within states, now internationalization might become the means of reducing them *among* states. An elementary form of world socialism might for the first time come into being.

It remains to be seen whether the challenge will be faced or the opportunity seized. This book is written in the hope that it may contribute at least to ensuring that the issue is better understood. With a more informed public opinion, it may be that the nations of the world will not be allowed to let this unique opportunity slip by.

E.L.

Note

This book was written before the author became a member of H.M. Government. The views it expresses are those of the author alone.

ABBREVIATIONS

ACC	Administrative Committee on Co-ordination
SCMSA	Sub-Committee on Marine Science and its Applications
ECAFE	(UN) Economic Commission for Asia and the Far East
ECOSOC	(UN) Economic and Social Council
EEC	Economic Commission for Europe
FAO	(UN) Food and Agriculture Organization
	Subsidiary Organs:
ACMRR	Advisory Committee on Marine Resources Research
COFI	Committee on Fisheries
GESAMP	Joint Group of Experts on the Scientific Aspects of Marine Pollution
GIPME	Global Investigation of Pollution in the Marine Environment
IAEA	International Atomic Energy Agency
ICES	International Council for the Exploration of the Sea
ICSU	International Council of Scientific Unions
	Subsidiary Organs
SCAR	Scientific Committee for Antarctic Research
SCOR	Scientific Committee on Oceanic Research
IGOSS	Integrated Global Ocean Station System
IHB	International Hydrographic Bureau
IMCO	Inter-Governmental Maritime Consultative Organization
IOC	International Oceanographic Commission
LEPOR	Long Term Expanded Programme of Ocean Research
UNESCO	United Nations Educational, Scientific and Cultural Organization
WHO	World Health Organization
WMO	World Meteorological Organization
	Subsidiary Organs:
CMM	Commission on Maritime Meteorology
PMAAP	Panel on Meteorological Aspects of Air Pollution

The Problem

CHAPTER 1

The Resources at Stake

The Traditional Resources of the Sea

Since the beginning of time, the bottom of the sea has been depicted, in myth and fable, as a place of untold riches. Davy Jones' locker has been painted as a store of fabulous wealth and wonder. Dreams have been built of the buried treasure of Atlantis and other legendary lands, lost, full fathom five, beneath the oceans, where bones become coral, eyes pearl, and all mortal dross, under the influence of the greeny deeps, suffers a miraculous sea change into something rich and strange.

It is only over the last century that it has begun to be realized that such stories are less fabulous than was once believed and that they may indeed be little more than the sober truth. The bottom of the ocean could bring forth all the riches long attributed to it, and more, far more than the most imaginative story teller had ever dreamed. This treasure may even, it appears, possess the magical quality that would have been dismissed as wild and fanciful imaginings in any fairy tale: the quality by which it grows and increases all the time, as if by an unseen hand, so that however much man takes away, he is always bound to find when he returns more than when he left. Like some new sorcerer's apprentice, and more fortunate than his luckless predecessor, he may discover that, however fast he fills his bucket from the sea floor, the good things he finds there not only will not be diminished but will continue to spread and increase much faster than he can pick them up.

The sea-bed covers more than 70 per cent of the earth's surface. It contains the earth's deepest valleys, some of the highest peaks and the largest mountain ranges, including volcanoes not always fully extinct. Its geological composition varies widely: from the continental crust areas of the shelf and the slope, which are essentially similar in formation and composition to the adjoining lands and contain many of the same resources, to the area of the deep-sea floor, which belongs to a different layer of the earth's primordial matter and is quite different in composition.

The upper, or *sialic*, layer of the earth's crust is confined to the continental areas and their immediate environment. It consists largely of granite, but also has sedimentary deposits laid down at later times, including many organic remains such as oil and gas. The *simatic* layer beneath this encircles the whole of the earth, including the floor of the deep sea, and consists largely of basic magmatic rocks, often covered with a thick sediment of silt, including the red clay that covers part of the oceans, sand, volcano dust, plankton, shells, teeth and bones of sharks and whales. This simatic layer probably contains plentiful quantities of minerals such as nickel, cobalt, platinum and copper. However, it is unlikely to contain the sedimentary rocks, in which gas, coal and other organic deposits are found. Here, therefore, the resources of greatest interest lie upon, rather than below, the bed.

It may be useful at the outset to define the terms which geologists use in describing these zones of the ocean floor, terms that will often recur in this book.

Closest to the shore, in effect an extension of the land mass beneath the water, is the area known geologically as the *continental shelf* (we shall discuss its legal definition in the next chapter). This slopes down relatively gently, usually at about 1/8th of one degree, from the low-water line of the coast to a point where the inclination becomes, usually quite suddenly, perceptibly greater: more than 3 degrees. This occurs normally at a depth of from 130 to 200 metres, but in exceptional cases at a point as shallow as 50 metres or as deep as 500 metres. The *breadth* of the shelf, because of the varying gradient, may range from one to 800 miles.

Beyond this, there is the *continental slope*, where the inclination is steeper: from 3 degrees to over 45 occasionally, though usually it is somewhere around 25 degrees. This goes out to a depth of about 1,500–2,500 metres. Because of the steep slope, it is a relatively narrow strip, usually 10 to 20 miles wide.

Next comes the *continental rise*, where there is still often a slight inclination, sloping down to somewhere between 3,500 and 5,000 metres in depth; there the bed flattens out. The rise is anything between 100 and 1,000 miles wide. Its surface is smooth and

Table 1

Basic Facts about the Ocean and Its Floor

Total surface of the ocean:	360 million km^2 (or 71% of the earth's surface)
Average depth of the ocean:	3,795 m
The area with depths between 4,000 and 5,000 m represents 36.6% of the ocean floor surface.	
Total volume of the ocean:	1,370 million km^3
Continental shelf (0–200 m)	7% of the ocean floor
Continental slope (290–2,440 m)	11% of the ocean floor
Area to 2,000-m isobath	16.3% of the ocean floor
Area to 3,000-m isobath	24.8% of the ocean floor
Continental margin	21% of the ocean floor
Abyssal plain (2,440–5,750 m)	79% of the ocean floor
Deep-sea trenches (more than 5,750)	3% of the ocean floor
The deepest trench is the Mariana Trench in the Pacific	11,304 m

covered with clastic sediments—that is, broken rocks of an older geological age. Its basic subsoil is, geologically, part of the deep ocean floor. Occasionally, it is sliced by very deep trenches, like the famous Norwegian Trench off the coast of Norway (though that is itself within the continental shelf).

Finally, there is the deep ocean floor, or abyssal plain, which is anywhere between 3,500 and 6,000 metres below the surface, with an average depth of just under 4,000 metres. It is by no means flat, however, being covered with seamounts, large mountain ranges and deep trenches. The deepest of these trenches, the Mariana Trench in the Pacific, goes down to a depth of 11,000 metres (more than 7 miles).

These three zones together—the shelf, the slope and the rise—are sometimes known as the *continental margin*. In fact the rise is, from geological and other points of view, better regarded as part of the deep ocean floor.

The geological character of these areas varies widely. Much of what we now call the shelf was dry land during the latter part of the Ice Age—that is, within the last 100,000 years. At that time, the sea sank two or three hundred metres because of the formation of glaciers, and subsequently rose again. It is largely for this reason that the shelf area is so rich in resources. The existence of plant and animal life in this period, and the weathering that the shelf underwent when the area was dry, were responsible for producing some of the organic materials beneath the sea which are now so valuable to us, especially oil and natural gas.

The resources of the ocean can be classified in a number of ways: according to their origin and method of formation; according to their chemical composition; according to the method of exploitation. Perhaps the simplest way is to divide them according to where they come from: the waters of the ocean; the floor of the sea; or beneath that floor.

The resources of the *waters* of the oceans[1] have, of course, been harvested for thousands of years—since fishing first began. Until recently, however, fishing was inefficient and unscientific, and concentrated in a few parts of the world that were not necessarily the most productive. Until the Second World War there was, for instance, very little fishing anywhere in the southern hemisphere, though some of the southern oceans, especially off the west coast of Latin America, are now recognized as the most productive waters in the world. There has been a huge increase in the value of production of fisheries during the last twenty-five years. Before 1939, the total world harvest of fish from the oceans was only 18 million tons a year. Today, approximately 65 million tons of fish a year are caught.

There has been some fluctuation in this increase. In the mid-sixties, production increased very little, largely because the catch of anchovies off Latin America ceased to grow. Atlantic cod and herring are now overfished, as are some other species. Already, therefore, the situation is changing radically. In the mid-fifties, virtually all fishing stocks outside the North Atlantic and the North Pacific were under-exploited, or not exploited at all, whereas now there are few stocks of the types of fish generally caught and marketed which are not heavily exploited, many by large fleets of long-range vessels capable of fishing anywhere in the world. The FAO estimates that the total world catch from

[1] This book is mainly concerned with the sea-*bed*, its future use and control, rather than the oceans, which are legally and economically quite distinct. But because the system for exploitation and control of the region may affect both, some discussion of problems concerning the waters as well is given.

conventional species could grow to over 100 million tons with a continued improvement in techniques. However, many of the most valuable stocks of fish are concentrated in waters fairly close to the coast, though outside territorial waters as traditionally defined. Many countries have therefore extended their fishing limits. That in turn is leading to increasingly bitter disputes about fishing (of which those between Britain and Iceland and Morocco and Spain are recent examples) and a race to carve up the waters of the sea. Ocean fishing is largely concentrated among developed countries with highly capitalized fleets, such as Japan, the United States and the Soviet Union. Developing coastal states extend their limits to preserve fish from these fleets. A confrontation is thus emerging between rich states and the poor, which see flourishing stocks of fish off their shores taken by the mechanized fleets of rich countries, yet may not themselves be capable of fishing those stocks fully.

The rapid increase in the total catch is the result of a number of improvements in techniques: more scientific study of marine stocks and their habits, better ways of locating fish, especially through the use of sonar, aircraft and satellites, improvements to fishing boats, nets and trawls, mechanized net-handling, fish pumps, factory ships for freezing, processing and storing fish, and more intensive fishing of some regions and of some types of fish little caught before, especially the tuna and the anchovy. Some of these improvements could increase the catch still further if they were more widely applied (for example to anchovy and sardines). But some have already caused over-fishing of certain species. One of the difficulties is that the methods of restriction traditionally used—banning the use of certain techniques or equipment—are technically irrational and may prevent more efficient and economical methods. Some believe that eventually only a restriction in the numbers engaged in fishing may ensure effective conservation, especially of certain species. But this will present obvious political problems in deciding whose fleets are to be cut out.

The most important single reason for the increase in the catch in recent years is the development of more efficient methods of tracking and following fish by fishing fleets. Fishing boats are larger and faster and can follow fish shoals more easily. The U.S. Fish and Wild Life Service seeks to trace the movement of tuna to help its fishing fleets. The Japanese fishing authorities prepare daily maps of the location of fish on the basis of reports from fleets.

In Britain, the Ministry of Agriculture, Fisheries and Food plots the movement of herring. Echo sounders, studies of ocean temperature and salinity, and other devices are used to map the presence and movement of fish. These expensive techniques are not of course evenly spread among countries and regions, with the result that there is a growing inequality in fishing power, which aggravates the political disputes. The differing fertility of waters, yielding from 400 pounds an acre in the fertile waters off the coast of Chile and Peru to 20 pounds off the coast of the United States, to an average of 1 pound an acre elsewhere, intensifies these problems. All fleets search the most favourable waters, whether they operate close to their own coasts or far away.

Another means of increasing the resources obtained from the seas, which has been much discussed in recent years, is by aquaculture, or fish farming: the deliberate cultivation of particular types of fish, with the object of securing a larger food harvest, just as animals are deliberately reared and fattened for food purposes on land. A considerable part of the effort of fishing at present is devoted to the search and capture of fish. The deliberate cultivation of fish for food purposes could reduce this cost: in fertilized fish ponds in Japan, production is several times as efficient as it is in the sea. Fish farming already occurs in the cultivation of such fish as oysters and clams. Aquaculture is now producing 6 per cent of all Japan's sea fisheries, and 15 per cent of its value. Japanese production of oysters, through intensive cultivation, has produced 50,000 pounds of fish an acre, against only a few pounds in many other places. In Scotland, plaice have been artificially reared in tanks, and already a million fish a year can be reared by this means. Cod and salmon are bred artificially and have even been trained to respond to the call of a signal, so that they can be caught easily when they reach the right age. Elsewhere, species have been transplanted into new areas to increase the total catch possible: striped bass, formerly found only in the Atlantic, have been transferred to the North Pacific, where they are now heavily fished. The intensification of these methods may increase the total number of fish caught,[1] but will not remove all the political problems over jurisdiction.

Another traditional type of marine resource from which much

[1] For a more detailed description of possible developments in fishing techniques and resources, see J. Bardach, *Harvest of the Sea* (London, Allen & Unwin: 1965); T. Loftas, *Wealth from the Oceans* (London, Dent: 1967).

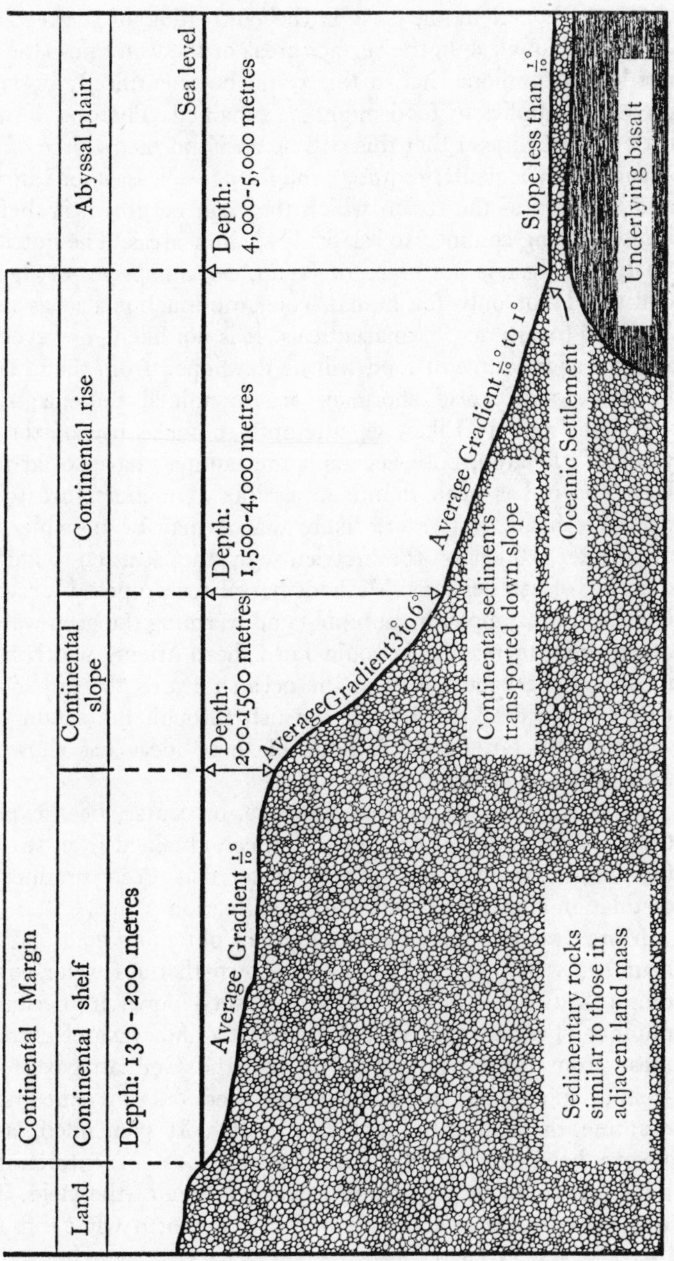

Diagrammatic profile of the Continental margin

Continental Margin

Land

Continental shelf

Continental slope

Continental rise

Abyssal plain

Depth: 130 - 200 metres

Sea level

Depth: 200-1500 metres

Depth: 1,500-4,000 metres

Depth: 4,000-5,000 metres

Average Gradient $\frac{1}{10}°$

Average Gradient 3° to 6°

Average Gradient $\frac{1}{10}°$ to 1°

Slope less than $\frac{1}{10}°$

Continental sediments transported down slope

Oceanic Settlement

Underlying basalt

Sedimentary rocks similar to those in adjacent land mass

has been hoped in the past is the cultivation of seaweed. The abundance of algae in the surface areas of the water near the coast has led to the hope that, if this could be scientifically harvested, plentiful supplies of food might be obtained. There is, however, little real likelihood that this will be an important source of food. Algae, like all plants, require sunlight as well as other nutrients, and the area of the sea in which they can be grown is therefore limited, being confined to relatively shallow areas. The nutritional quality of algae is not high. In Japan, certain species of algae are cultivated not only for human consumption but also as animal feed and for use in pharmaceuticals. It is not likely, however, that a new major source of food will be developed from them, largely because today world shortages are in animal proteins, not in vegetable matter. Likewise, attempts to make use of the food value of plankton, both vegetable and animal, have so far failed. Here too it has been found impossible to manufacture food in palatable form. It is more likely that it may be possible to redistribute artificially the nutrients in the sea, on which the quantity of fish depends. Methods have been proposed for achieving this, for example, by pumping and warming the deep waters to cause them to rise. This could raise the nutrients which mainly inhabit the deeper waters of the ocean towards the upper areas, where alone they can benefit the fish through the action of the sunlight. So far, however, none of these ideas has shown any signs of being economic.

Mineral wealth of the seas has also, of course, been exploited traditionally. Salt, for example, has been obtained from sea water for thousands of years.[1] Fresh water has been produced by distillation and more recently by desalination plants.

More dramatic has been the hope of obtaining gold and silver from sea water. It has long been known that sea water contains minute quantities of these metals washed down from rivers. A number of attempts have been made to find ways of extracting these from the ocean. In the 1920s, the German government, seeking money to pay war debts, fitted out an expedition to examine these possibilities scientifically. It concluded, as has always been concluded since, that the cost of extracting the minerals would be far greater than the value of the yield. There are, however, other minerals which it is worth while extracting

[1] Production of salt from sea water is worth about $200 million at present, and is the most valuable resource obtained from the sea after oil, gas and iron ore.

from sea water: bromine, magnesium, magnesium compounds, and heavy water are all now being manufactured from sea water in the United States and elsewhere. An estimated value of minerals produced from sea water is shown in Table 2.

Another possibility exists of drawing wealth from sea water. It has recently been discovered that at the bottom of certain seas there exist hot brine pools that are far richer in minerals than the oceans as a whole. They appear to result from hydro-thermal brines from within the earth forcing their way up through sea-bed fissures and faults. At the bottom of the Red Sea, for example, such a brine pool has been found which contains huge concentrations of iron, manganese, zinc, lead, copper, silver and gold, at concentrations 1,000 to 5,000 times greater than that normal in sea water. Even higher concentrations are found in the bottom mud of the sea in such areas. One relatively small pool in the Red Sea, Atlantic Deep II, is thought, from a sampling of the bottom sediment, to contain in its upper 10 metres of mud about $2,300 million worth of copper, lead, zinc, silver and gold.[1] Sediments rich in minerals have been reported also from a sea-bed volcano off the coast of Indonesia. It is not yet possible to exploit these minerals because of the depth of the water, but with the advances now being made in underwater technology, which we shall consider shortly, it may not be much longer before they can be extracted economically. Here, the question will arise (it has already occurred in the Red Sea): who has a right to the minerals so discovered?

The Sea-bed

On the sea floor itself other discoveries potentially far more important economically are being made than those outlined so far. In fact, it is now realized that the greater part of the riches of the earth may be situated here.

The resources on the sea floor belong mainly to three categories: firstly, the basic natural materials of the floor, whether in the continental crust or beyond; next, the placers, deposits that have been washed from rivers and elsewhere and laid on various parts of the sea-bed; and lastly, the precipitates, which have been formed gradually within the sea water or on the bed itself.

[1] See UN Doc. E/C7/2: Add. 5, January 19, 1971, Report of the UN Secretary-General, *Marine Mineral Resources.*

Table 2

Minerals from the Sea: annual production
(Value in millions of U.S. dollars)

From sea water	
Salt	172
Magnesium metal	75
Fresh water	51
Bromine	45
Magnesium compounds	41
Heavy water (D_2O)	27
Others (potassium, calcium salts, sodium sulphate)	1
Total value from sea water	412
From sea floor (surface deposits)	
Sand and gravel	100
Shell	30
Tin	24
Heavy mineral sands (ilmenite, rutile, zircon, garnets, etc.)	13
Diamonds	9
Iron sands	4
Total value of surface deposits	180
From sea floor (sub-surface deposits)	
Oil and gas	6,100
Sulphur	26
Coal	335
Iron ore	17
Total value of sub-surface deposits	6,478
TOTAL	$7,070

Source: Marine Science Affairs—Selecting Priority Programmes, Annual Report of the President to the Congress on Marine Resources and Engineering Development, U.S. Government Printing Office, April, 1970 (UN Doc. E/4973 of April 26, 1971).

In the first category, the most important are sand and gravel, which are dredged from the bottom of the sea for building and other purposes in many parts of the world. In terms of value this is still probably the most important single product of the sea-bed; yield is thought to be worth around $100 million a year. Shells, iron-sands and heavy mineral sands are also dredged to a total value of around $50 million a year.[1]

Of the placer deposits, tin has been mined from the bottom of the sea off the coasts of Malaysia, Thailand and Indonesia for many years. Diamonds have been obtained from the sea floor off South West Africa. Many of these minerals can probably be profitably exploited only within the continental shelf and in other areas close to the coast.

It is the resources of the third kind, the precipitates, that are now proving to be by far the most important of the sea floor deposits. So far, the only one that has been exploited economically is phosphorite. This is found in the continental shelf and slope, and occasionally on the deep ocean floor. Its importance is that phosphates can be extracted from it for the manufacture of fertilizer and chemicals. It is formed from decayed animal and plant remains, washed down from rivers or carried upwards by rising currents from ocean floor sediments to shallower areas. It is found in the form of nodules, slabs, pellets and coatings of rock, as well as in phosphate sands and rocks. It occurs particularly where little other sediment overlies it—that is, where rainfall is low and few large rivers wash into the sea, or opposite dry desert areas—for example, off the west coast of the Americas and North West Africa. Most of the deposits found so far contain only 20 to 30 per cent of phosphate, which means that extensive chemical upgrading or other processing is necessary if production is to be competitive with phosphates obtained from the land. Because at present superior phosphates are cheaply and easily available in many land areas, production from the sea-bed is not always economic. (An attempt at off-shore mining in California was abandoned as unprofitable in 1965.) But demand for phosphates is steadily rising, and marine exploitation could become increasingly important for a country having few land sources, or where transport costs would otherwise be heavy.

[1] See Table 2, p. 12. See also G. M. Fye, *et al.*, *Ocean Science and Marine Resources*, in E. A. Gullion, ed., *The Resources of the Sea* (Englewood Cliffs, N.J.: Prentice-Hall, 1968, p. 33).

The most important of the precipitates, however, is of a different nature. In 1872, the British survey ship *Challenger* set out on four years of exploration of the ocean bottom. It took sample dredges in many different parts of the oceans and made many important discoveries about the nature of the oceans and the ocean floors. Its findings were eventually published in thirty-two volumes, and effectively initiated the science of oceanology.

Perhaps the most important discovery was one whose significance was not at first realized. This was the discovery in the Pacific, at the bottom of the ocean, of small nodules of metal. Analysis revealed that these were predominantly composed of manganese, and they have come to be known as manganese nodules. About thirty years later, another expedition, in the *Albatross*, discovered that these nodules covered an area of the Pacific as large as the United States. Since that time it has been discovered that nodules are widely scattered over the ocean floor, predominantly but not exclusively in deeper waters, in the Pacific, Atlantic and Indian oceans.

The nodules are on average about 5 centimetres in diameter, and vary from one to 20 centimetres. Rarely one is found with the diameter of a metre, and one was recovered with a weight of 750 kilograms. They are irregular in shape, nobbly and round, like potatoes. They are brown or black in colour, and usually soft, porous and crumbly. They are often formed round a nucleus, such as a shark's tooth, or a whale's earbone, or a stone. Coatings of manganese and other substances are also found on rocks and sea-mounts, sometimes from 2 to 10 centimetres thick. Finally, there are also small grains of manganese dioxide about half a millimetre across, within the clays and oozes of the ocean bottom.

If manganese was the only substance in the nodules, they might not be of great importance, as manganese is relatively common on land. But, though manganese is the largest single constituent, the nodules contain a number of other important metals, especially nickel, copper, cobalt, molybdenum, aluminium, iron, and in enormous concentrations. The proportion of the different metals in each nodule varies in different regions of the ocean.[1]

After the Second World War a far more systematic survey of manganese nodules was undertaken, especially in the Pacific.

[1] For a more detailed analysis see J. L. Mero, *The Mineral Resources of the Sea* (Elsevier, Amsterdam: 1965).

Undersea photography, a new technique, revealed their presence and density far more clearly than before. It began to be realized that in some areas they were thickly concentrated—in some places at perhaps as much as 100,000 tons per square mile. The total volumes of the metals they contained were therefore enormous.

The estimates vary widely. In 1958, H. W. Menard and C. J. Shipek published an article in *Nature*, on 'Surface Concentrations of Manganese Nodules', in which they estimated a total of 100 million tons for the south-west Pacific alone. In 1958 to 1959, a group of U.S. scientists from the University of California made some estimates of the total volumes of nodules in the entire Pacific Ocean, based on the size of concentrations revealed by undersea photography.[1] This study gave the figure of about 1,700 billion (i.e. thousand million) tons for the Pacific as a whole. In 1961, two Soviet scientists published another estimate for the Pacific, which was only about a twentieth as large, though still enormous. Other studies give still smaller figures. There is a great deal of guesswork in such estimates, and it is possible that many are inflated. The important point is that it is everywhere conceded that, whatever the true figure, the wealth contained in the nodules is unbelievably vast.

Such quantities imply that there are enormous amounts of the individual metals within the nodules. According to a recent UN study, the nodules may contain 16.4 billion tons of nickel, 8 billion tons of copper and 8.8 billion tons of cobalt.[2] The importance of the resources can be made clear by comparing them to the amounts of the same resources known to exist on land. If the calculations of the California study quoted above were confirmed, this would mean that the nodules of the Pacific Ocean alone would contain about 358 billion tons of manganese, equivalent to reserves for 400,000 years at the 1960 world rate of consumption, compared with known land reserves for only 100 years; 43 billion tons of aluminium, equal to reserves for 20,000 years, compared to known land reserves for only 100 years; 14.7 billion tons of nickel, equivalent to reserves for 150,000 years, compared to reserves to last 100 years on land; 7.9 billion tons of copper equivalent to reserves for 6,000 years, compared to reserves of only 40 years on

[1] See J. L. Mero, *The Finding and Processing of Deep-sea Manganese Nodules* (The University of California Press, Berkeley, Cal.: 1959).

[2] Report of the UN Secretary-General, *Marine Mineral Resources*, January 13, 1971 (E/CM.20DD5).

land; 5.2 billion tons of cobalt, equal to reserves for 200,000 years, compared to reserves for 40 years on land. The nodules may contain 207 billion tons of iron; 10 billion tons of titanium; 25 billion tons of magnesium; 1.3 billion tons of lead, and so on. These are figures for the Pacific alone. With the amount of metals in the nodules in the Atlantic and Indian oceans, all these figures, of course, become still larger.[1]

A further fact concerning the nodules is even more astonishing. It is now realized that the nodules are forming all the time, and at a stupendous rate—in the Pacific alone, perhaps at 10 million tons a year. A naval shell of a type made in the Second World War, found off San Diego, had a coating of iron-manganese oxides about 1.5 centimetres thick, formed in little more than twenty years (though the rate of formation is slower farther out at sea). Accretion in the shallower waters and on the continental shelf apparently occurs at a rate of 0.01 to 1.0 millimetres a year. The nodules are apparently formed and enlarged by precipitation from the water and by particle agglomeration. It has been suggested that, when manganese and iron in the sea water reach saturation point, they form colloidal particles, which are electrically charged and attract cobalt, nickel, copper, molybdenum, zinc, lead and other metals. At the same time, particles of minerals and organic debris, settling through the water, attract manganese, iron and other minerals. On reaching the sea floor, the colloidal particles are attracted to protruding objects, such as stones and bones, which may act as better conductors of electricity.

Such huge supplies of minerals are obviously potentially of enormous importance. How quickly they become actually important depends on how cheaply the metal itself can be made commercially available. This depends on the rate of advance in two techniques. The first is that of obtaining nodules from the sea-bed. This is unlikely to prove particularly difficult. Because the nodules lie mainly on the surface of the sea-bed, there is no

[1] It is of course not the total quantity of nodules but the amount that are of a quality and location to be fairly readily exploitable which most matters in the immediate future. According to a recent report, 'several industrial corporations in the USA and Japan have reportedly taken several thousand samples and have used under-way real time television to investigate vast areas of the abyssal floor. Exploration to date has shown that surficial nodules of exceptionally hight content of copper, cobalt, nickel, manganese and other metals are likely to be present in sufficient quantities at depths over 3,600 metres'. (UN Doc. E/4973, *Mineral Resources of the Sea.*)

inherent difficulty in picking them up by mechanical means (though the highest concentrations are at considerable depths). In 1970, a U.S. mining company demonstrated an airlift/suction hydraulic drain, designed for gathering up nodules at depths of 750 to 915 metres. In 1971 a U.S.–German consortium announced commercial recovery of nodules from the Pacific off Hawaii, and said that it planned to start full-scale deep-sea mining in the near future.[1] Far harder is likely to be the extraction of the different metals after discovery, especially because of the presence of silicon in the nodules. Much work, however, has been done on this recently. The scientists of the University of California in the study mentioned earlier concluded: 'Many of the industrially important materials can be produced at 50–75% of the present cost of producing the metals from land deposits'.[2] According to one study, the cost of extraction of the minerals would be only $25 a ton for manganese, cobalt, nickel and copper.[3] It is believed that hydraulic dredges, operating from the surface, would be able to draw up 4,000 tons of nodules a day at a profit of $7 a ton, and that some of the metals would be 25 to 50 per cent cheaper than at present.[4] Later studies by industrial firms have, it is claimed, confirmed these results. As a number of mining firms have been spending millions of dollars to develop the necessary equipment and methods, it seems that those who have most experience in this field are confident that the problems can be overcome relatively easily.[5] It will therefore probably take only a few years before these minerals are exploited commercially.

The exploitation of nodules has a number of significant advantages for large mining companies, and for the state that owns them. There need be no expensive mining or drilling. The material is already available on the surface of the sea floor and

[1] For the most recent position, see Chapter 13 below.

[2] J. L. Mero, *The Mineral Resources of the Sea* (Elsevier, Amsterdam: 1965), pp. 274–5.

[3] J. L. Mero, *Finding and Processing of Deep-sea Manganese Nodules* (The University of California Press, Berkeley, Cal.: 1959).

[4] J. L. Mero, *The Mineral Resources of the Sea, op. cit.*

[5] *Ibid.*, p. 275. A 1971 UN report stated that 'it is reported that in the USA investigations were carried out of several dozen known hydrometallurgical processes and then experiments with a few highly promising processes, particularly solvent extraction by acids and bases and separation of the metals by ion exchange or precipitation. A pilot plant will be built to further test and develop hydrometallurgical processes' (E/4973, p. 34).

can be fairly simply obtained; there are no costs for exploration, for large-scale construction, or for mineshafts and tunnels or other engineering works. Large expenditure in housing mine workers, in the building of access roads or railways, or in continually developing mines which are required on land is also avoided. There are at present no royalties to be paid to governments, nor any political risks. A large proportion of the material can be sold compared with only 2 per cent which can be extracted from the ores mined on land. Once the basic equipment has been developed and appropriate ships constructed, they can be used according to need and wherever it is thought profitable to mine nodules. Finally, not only are the basic raw materials never depleted, as on land, but they accumulate even while they are being exploited. These are spectacular advantages.

However, the very ease and speed of exploitation of nodules could present enormous dangers to those countries and populations at present dependent on land production of the same minerals. With rapidly increasing supplies, prices could decline sharply. And if sea-bed methods proved much cheaper and more economical, those countries and populations might lose their markets altogether, and with them their chief means of subsistence and source of foreign exchange.[1] It is thus not surprising that some leaders of the developing world, aware of this problem, have begun to look with apprehension at the possible effects on their economies if these minerals should appear on world markets. To them, as to others, the question immediately arose: if the economic potential of these new materials is confirmed, who will have the right to exploit them; and on what terms? Or, put more simply still, whose are they?

The Subsoil

Although manganese nodules may eventually prove to be the most important of the riches of the sea-bed, it is when we turn to

[1] 'If sufficiently low-cost production and refining methods are developed to permit large-scale nodule exploitation, sub-sea production could conceivably satisfy a large part of the world's need for each of the component metals, particularly copper, nickel, cobalt and manganese. . . . A single exploitation operation including an efficient processing plant could furnish 5.3 per cent of present world manganese consumption, 3.4 per cent of nickel consumption 14.1 per cent of cobalt and 0.2 per cent of copper.' *Mineral Resources of the Sea*, a report for the UN Secretary-General by F. Wang, E/4973, April 26, 1971, p. 35.

the resources below it that we discover those that are at present of the greatest economic significance. Some resources beneath the sea have been exploited for many years. Coal and iron ore mined from beneath the sea through shafts sunk on land in England, Japan, Newfoundland and Finland: about 30 per cent of Japan's coal now comes from beneath the sea. Salt and sulphur are obtained from the caps of salt domes in the Gulf of Mexico. Fresh water is pumped from beneath the continental shelf to supply many communities on the south-east coast of the United States.

Here, however, we are concerned mainly with oil and gas. Their extraction is, of course, economically feasible. They have been produced from beneath the sea for over a quarter of a century, during which time underwater operation has shown a spectacular growth.

Some oil was taken from wells in shallow waters off the coast of California in the last century. The first under-water oil drilling took place in Venezuela in 1923. However, to all intents and purposes off-shore drilling began only at the close of the Second World War, when large new reserves of oil were found beneath the Gulf of Mexico. There off-shore production began on a large scale, though usually in comparatively shallow water, less than 30 metres in depth, and remained at first only a small proportion of the total. In 1950 it was only about 3 per cent; even in 1960 it was still under 10 per cent. Today it is over a fifth of all oil production and is expected to be over a third by the end of this decade. By the end of the century the greater part of oil production may be from beneath the oceans. There are already indications that there is more petroleum beneath the sea than on land. Production from off-shore sources increased sixfold between 1960 and 1969 and off-shore expenditure is increasing at present by 18 per cent per year.

Similar developments are affecting production of gas. Indeed, the rate of growth is in some ways even more spectacular. Because gas and oil are both hydrocarbons derived from similar organic material, they are often found in similar locations. A number of the areas where under-sea oil has been found in recent years, in the North Sea, in the Persian Gulf and elsewhere, have also proved to be large sources of natural gas. Some of the undersea gas fields found, such as the Groningen field off the Netherlands in the North Sea, are among the largest gas fields known anywhere:

Table 3

The Main Sea-bed Areas

Continental Shelf

Width

Average: 40–60 miles
Range: less than 1 mile to over 750 miles

Depth (*outer edge*)

Average: 436 feet (133 metres)
Range: 164–1,804 feet
General range: 436–656 feet (130–200 metres)

Gradient

Average: 1:600 (0.1°)

Continental Slope

Width

Average: 10–20 miles
Range: 9.3–50 miles

Depth (*outer edge*)

Average: 5,998.4 feet (1,830 metres or 1,000 fathoms)
Range: 3,280–16,400 feet (1,000–5,000 metres)

Gradient

Common range: 2°–6°
Average: about 1:14 (4°)

Continental Rise

Width—Range: May be as much as 620 miles
Depth—Range: 4,920–16,400 feet (1,500–5,000 metres)
Gradient—Range: Less than 1:40 (about 1.5°) down to 1:1,000
Thickness—May be 0.6–6.2 miles (1–10 kms)

Percentage of Sea Floor at Various Depths

According to Svendrup, Johnson and Fleming, *The Oceans*, 1942, the percentage of sea floor between various depths is as follows:

Depth Interval (metres)	*Individual Zone*	*Cumulative Total*
0– 200	7.6	7.6
200–1,000	4.3	11.9
1,000–2,000	4.2	16.1
2,000–3,000	6.8	22.9

the Groningen field alone is said to contain 40 billion cubic feet of natural gas. U.S. off-shore gas reserves are estimated at 200 trillion cubic feet. Off-shore gas production in the United States doubled between 1960 and 1965 to nearly 1,000 million cubic feet, and has more than doubled again since. Total world production of natural gas is expected to triple within the next decade, and a substantial part of these supplies will be represented by undersea gas.

So far, most off-shore production of both oil and gas has been within the continental shelf itself, mainly in depths of under 100 metres. The deposits exploited have largely been in areas adjoining others known on land: the Gulf of Mexico, the Persian Gulf, and off the coast of California. But as exploration intensifies, as the demand for oil increases, and as the technology of off-shore production improves, exploitation is taking place at greater depths. Although in 1960 most off-shore production was in depths of 30 metres or less, today much takes place at 100 metres and more. Exploration or 'wild-cat' wells today are often at depths up to 200 metres, and have been sunk at nearly 500 metres (for example, off the Californian coast). During geophysical reconnaissance work, shallow core holes have been drilled in waters as deep as 1,500 metres. Production may even take place at 500 metres, well beyond the average physical continental shelf, in the next few years.[1]

It is known that oil exists well beyond the edge of the continental shelf, at least to the edge of the continental margin. A drilling in August, 1968, part of the Deep Sea Drilling Project sponsored by the National Science Foundation in the United States, found traces of oil and gas at 3,500 metres in the Gulf of Mexico. At least in relatively small oceanic basins not too far from the land, there are good chances that oil may be found even beneath the abyssal plain of the deep ocean. One study has concluded: 'Because of man's limited knowledge, no definite seaward limit for the existence of petroleum deposits can be inferred at this time, and it is not impossible that small portions of the abyssal floor and oceanic trenches may have some potential.'[2]

[1] 'There is reason to believe that commercial exploitation of petroleum will be possible in 457 metres (1,500 ft.) of water by 1975. The technological limit would extend much further off-shore, probably as far as 1,800 metres (6,000ft.) water depths within the next decade.' *Mineral Resources of the Sea* (E/4973).

[2] Report of the UN Secretary-General, *Mineral Resources of the Sea*, E/4680, June 2, 1969, p. 19.

As drilling takes place at greater depths, the cost of production increases sharply. Off-shore expenditure, even at existing depths, is 3 to 5 times greater than that on land, and development costs rise rapidly as the depth of the water and distance from the land increases. According to some, the cost doubles as the depth increases from 30 to 180 metres, and more than doubles again in moving out to 300 metres. In areas of strong tides or ice floes, the additional costs would be greater still.

Against these higher costs of production, exploitation of off-shore oil has important advantages both for companies and for governments. They can substantially reduce the balance-of-payments cost of oil, an important factor in itself. Off-shore production in home waters usually involves smaller royalty and tax payments than that from land sources abroad: the very high payments recently demanded by the oil-producing governments further increase this advantage. Production is not subject to political pressures or risks. There are obvious strategic advantages. Off-shore production is also often closer to the ultimate markets and, since the cost of transport is a large part of the total cost of oil, this can be an important consideration.

Moreover the demand for oil and other forms of energy is increasing so fast that all possible sources must now be exploited. Total demand for oil is expected to double in this decade. This means that, despite the high costs, there is likely to be increasing production from the slope areas (in other words, outside the areas traditionally regarded as within national jurisdiction), especially where large productive fields exist in convenient locations. The sea-bed is vital to close the energy gap that now yawns so wide.

For all these reasons exploration of oil deposits in off-shore areas will probably become more intensive than ever. Exploitation will take place at greater and greater depths. Increasingly, this means, exploration and even production is spreading out into areas which are geologically well beyond the continental shelf, and so perhaps beyond national jurisdiction. This gives rise to the same problems as have occurred in the case of nodules: to whom do these resources belong; and who will have the right to exploit them, under whose authority and under what conditions?

The Technology of the Sea-bed

How soon this new wealth will be exploited depends on how

quickly the necessary technology develops. Dramatic advances in sea-bed technology have taken place in the last ten years. They have already made possible the exploitation of oil at depths which once would have been quite unthinkable.

First, improvements have been made in methods of surveying the ocean bed. Sophisticated echo sounders send acoustic signals down to the sea bottom which, on their return, are amplified and converted by a recorder to give a visual profile of the sea floor. Such soundings can not only record the depth but can also show whether the bottom is rocky, sandy or muddy. This picture can be filled out by 'side-scanning' sonar, which gives similar pictures in a lateral direction, and so in effect enables men to see in opaque waters for up to half a kilometre. In addition, underwater cameras can today take stereophotographs of the sea bottom at considerable depths.

In recent years, these methods have been supplemented by seismic and sonic profiling of the rocks and deposits *beneath* the sea-bed. It has been found that sonic pulses can penetrate for short distances into the mud of the sea floor, and the form of the echoes can be used to indicate its stratification. A transducer sends sounds of varying frequencies and types down to the bottom, and a complex acoustic receiving unit, with sensitive crystal detectors, records the returning sounds and processes them to present them as continuous profiles of the stratification beneath the sea-bed. Many ingenious types of machine, making use of different kinds of sound—the electric sparker, the boomer, the gas exploder, the pneumatic sounder and vibrosis transducer—have now been developed. Some of these can produce sub-bottom penetration down to 6 or 7,000 metres below the sea-bed. Since knowledge of the geological structure at this level is often necessary in discovering new sources of oil, these techniques have acquired a special importance in under-water surveying.

These methods are supplemented by other techniques. By recording the travel time, and so the distances, of the shock waves from a high explosive or other source of energy (seismic refraction and reflection) it is possible to measure the configuration of rock structures. Highly sensitive magnetometers, carried in surface vessels or submersibles, are used to trace the presence of large crystal structures, faults, rock types and other features of the sea floor. Gravimeters, which record gravity, encased in watertight diving chambers, are moved along the sea floor with remote control and recording devices. Variations in gravity are then used to interpret the structures and types of rock beneath the sea-bed. Aircraft and satellites, using sensors, can measure sea temperature, currents, ice, sea level, pollution and other features, and aerial photography can penetrate clear, shallow water and so record the sea floor. Underwater cameras and television give pictures of the bottom of the sea, showing the distribution of rocks, nodules or other features. New methods of core-drilling have been developed which enable large samples to be taken up from considerable depths down in the sea-bed. Drilling has sometimes taken place to 9,000 metres, in water which is itself several hundred metres deep. Conversely, some ships, such as the *Glomar Challenger*, have drilled to nearly 1,000 metres, in waters 6,000 metres deep.

All these methods are today combined in modern surveying expeditions. Integrated shipboard systems, using magnetometer, gravimeter, echo sounders, seismic reflection, core drilling, geochemical sampling and underwater sensors to detect oil seeps, are carefully co-ordinated with measurements of the ship's position at different times. The data are presented on analogue recorders and magnetic tape for computer processing and correlation. They are used to provide comprehensive pictures of the probable configuration of the sea bottom and sub-surface deposits.

Through such improvements in surveying techniques, it has been possible to explore resources situated much below the sea-bed and far beyond the edge of the continental shelf as usually defined. This in turn has demanded development of the technology for exploiting those resources once discovered. Here the first advance required has been in the construction of drilling rigs and platforms. The first off-shore rigs were of a type similar to those used on land, placed on a platform fixed to the sea floor, usually by steel piles. In recent years, floating platforms have been developed which can be moved about on the water, and are sometimes semi-submersible. On some of these, drilling takes place through a well in the middle of the vessel. A U.S. firm has constructed a drilling platform for production in over 200 metres of water. It is expected that platforms and rigs will shortly be in operation capable of operating for production purposes at depths of 1,000 metres. Drilling platforms of this kind are of course highly expensive pieces of equipment, costing up to £50 million each.

There have also been improvements in the drills themselves. Powerful diesel-electric engines are used to drive them, and this makes possible the drilling of multi-directional wells. One of the main difficulties of previous times, that of securing re-entry to the same hole once the bit was removed, has now been overcome. Techniques of drilling at angles of 65 degrees have been developed, making it possible to tap reservoirs a mile from the well head. Drill-testing equipment has been developed, using microwave systems, which transmit the data received from the well to headquarters on shore.

Automatic systems are used to operate the wells by remote control, through acoustic signals and electro-hydraulic systems. For example,

An American deep-water production system now under construction involves a combination of deep-water fixed platforms, and a 'satellite production system' (SPS) which could be installed on the sea floor in as much as 610 metres (2,000 feet) of water and connected to the platform by pipelines. . . . A typical SPS unit would contain up to forty subsea wells, including equipment for manifolding and production control, gas injection, well maintenance and a pollution control system. The SPS operations are remotely controlled from a nearby platform by means of coded electronic signals sent via submarine cables. The SPS unit is so designed that all equipment on the template can be serviced by a manipulator work unit lowered from the surface and moved around the structure on rails.[1]

Complete underwater systems designed to exploit depths well over 200 metres in depth are now being designed. A development reminiscent of science fiction is the use of underwater robots controlled by electric cables from the surface (Mobots), which can service the wellheads beneath the sea in place of human divers, who cannot descend to great depths directly from the surface.

There are also under-sea cars. A sea-bed vehicle constructed in 1970 in the United Kingdom (the Cammell Laird) can support two operators and three divers on the sea floor to depths of 182

[1] UN Doc. E/4973, *Mineral Resources of the Sea*. This report describes another system in which 'each wellhead has two miniaturised long-duration radioisotope generators which power the electromechanically actuated valves without interruption. Production controls for these wells are actuated and precisely regulated by coded acoustic signals sent out from a central production console located on a fixed platform about 1.5 kilometres from the wells. The two-way acoustic telemetry system transmits operational status reports upon request from the production console and automatically transmits any unscheduled operational changes or malfunctions. Most sub-sea equipment used is designed to be set up from the sea-surface without diver assistance.'

metres (600 feet) and perform work such as pile-driving, pipeline inspection, trenching, pipeline burying, salvage, and maintenance. The vehicle travels on large wheels along the sea-bed, and power and life-support services are fed from a surface support vessel.

There have been important developments in other fields. The technology of underwater vessels has been rapidly advanced in recent years. The French balloonist Auguste Piccard first developed a large-scale submersible vehicle, later taken over by the U.S. Navy. His son designed the bathyscaph *Trieste*, which in 1960 dived down to the deepest part of the ocean, 11,000 metres beneath the surface. This vehicle was not entirely satisfactory, however, because it required surface support, and could not easily be manoeuvred. In 1964, the U.S. Oceanographic Institute developed the *Alvin*, which was much more manoeuvrable than previous submersibles, and operated safely at 2,000 metres. This was successfully used to locate the hydrogen bomb accidentally dropped in the Mediterranean as a result of a collision between aircraft in 1966. Another U.S. vehicle, the *Aluminaut*, constructed of aluminium, and originally intended to operate at 5,000 metres, was able to transport six men for 100 miles at a depth of around 2,000 metres. Since that time, considerable advances have been made through the use of glass to provide hulls capable of withstanding the heavy pressures of deep water. Westinghouse, the American corporation, now advertises commercially a submersible, the *Deepstar* 20,000, which is capable of exploring more than 98 per cent of the ocean bed, with forward- and side-looking sonar, low-light-level television cameras, radar transponder, movie camera, telecommunication equipment and two manipulators, for exploration, salvage construction and rescue.

There have also been very rapid advances in the technology of dredging, which will be important in the exploitation of nodules. In 1970, a pilot-scale airlift hydraulic dredge system was successfully tested in the Blake Plateau off Georgia (an area known to be rich in nodules) in depths of 800 to 900 metres. A high-strength flexible dredge pipe over 760 metres long, made up of bolted 10-inch segments, is dropped from a derrick platform over a central well within the ship, which moves at one to three knots while it operates. A television camera on the dredge head monitors its performance and helps in eliminating unwanted sea-floor material and directing the nodules into the dredge pipe. Pump units along the pipe are used to move the fluid–solid mixture upwards. On the ship a device separates the nodules from the water and other materials, while a conveyor distributes the nodules to the ship storage. The test showed that the pipe could withstand pressures and that the dredge had sufficient power to lift several hundred tons of nodules through the dredge pipe. On the basis of the test, an operational mining rig was constructed which could mine 4,000 tons of nodules a day from 6,000 metres. The company concerned, Deepsea Ventures, believed that full-scale mining could begin by 1974 and that refined minerals could reach the market in 1975. It is also thought possible that the same system might be used to bring about up-welling from nutrient-rich deep waters and so increase fish life near the surface. Another test carried out in the Pacific by a Japanese firm demonstrated the ability of a continuous line bucket system to dredge in waters up to 3,700 metres in depth. An improved system capable of recovering 500 tons of nodules a day from depths of 5,400 metres is now being designed.[1] Howard Hughes is investing in similar systems.

Lastly, an important advance is the development of techniques by which men may live for considerable periods below the surface of the sea. During the Second World War the development of the aqualung made it possible for divers to descend to considerable depths independently of any contact with the surface. But it remained impossible for them to go below about 70 metres, because physiological changes in the diver induced the bends, a form of intoxication (during which he might suddenly decide to abandon all his equipment), or even death. Recently, use of the technique of saturation diving, which has been developed by the U.S. and British navies, has made it possible for men to descend to depths of 500 metres.[2] The diver is placed in an artificial atmosphere at depths well below the surface, so that his physiology becomes sufficiently acclimatized. It is then possible for him to descend to greater depths. Once acclimatized at 70 metres, he may descend down to 150, and from 150 to 300. By these means, a number of men have spent periods of up to a month at 200 metres, with limited excursions to greater depths. Such techniques have

[1] The Japanese Ministry of Transporattion has recently announced plans to design and build a manganese nodule mining ship.

[2] In 1970, volunteer members of the Royal Naval Scientific Services, United Kingdom, stayed 10 hours at a world record simulated 'dry' diving depth of 457 metres (1,500 feet). In November, 1970, two divers from the French Oceanographic Agency (COMEX), broke this record in a simulated 'dry' dive to 520 metres (1,706 feet). UN Doc. E/4973, *Mineral Resources of the Sea*.

made it possible to place permanent manned stations on the sea-bed at depths of 600 or even 1,000 feet.[1]

Experiments involving the use of fixed or mobile sea-floor habitats have been conducted in many parts of the world, and it is expected that habitat technology may soon find its full application in off-shore petroleum work. Development of the ability to work and live in permanent sea-floor bases for prolonged periods under considerable hydrostatic pressure is expected to be an integral part of the new technology that will evolve as the petroleum industry moves into deeper waters.

Developments are thus taking place which, twenty or thirty years ago, would have been thought unimaginable outside the pages of H. G. Wells. Advances may be made faster still in the future. For more money is today being devoted to such research, by governments as well as by companies, than ever before. In the United States alone, it is said that more than $5 billion a year is now spent on research relating to undersea technology, either by the U.S. government or by U.S. companies. It may be that the technical problems that remain, such as the economic separation of the metals of the manganese nodules, may be overcome within this decade. If so, it may make possible the production of huge supplies of metals and other resources, and at a far lower cost than from land sources today.

This will raise huge political and legal problems and could be the cause of bitter international conflict. A struggle has already begun for control of these immensely valuable resources. We must next look at the legal problems that have arisen over rights in the ocean areas.

[1] For further descriptions, see G. M. Fye *et al.*, 'Ocean Science and Marine Resources', in E. A. Gullion, ed., *Uses of the Seas* (Englewood Cliffs, N.J.: Prentice-Hall, 1967), p. 28.

CHAPTER 2

The Law in Dispute

The Surface of the Water

The discovery of the enormous wealth beneath the sea, perhaps many times greater than that known on land, raises the obvious question: to whom does it belong? Let us consider first what traditional law has to say on this question.

In the last chapter we described the various areas of the sea floor, as geologists distinguish them. We must now look at the distinctions and definitions made by international law. These have concerned both the surface of the water and the sea-bed below.

On the surface there exists, first, adjoining the coast, a strip of water known as the *territorial sea* or territorial waters. This area is, to all intents and purposes, a part of the territory of the coastal state. That is, that state, while it remains under an obligation to allow freedom of peaceful and 'innocent' passage, can exercise there all its normal rights of jurisdiction and sovereignty. It can issue laws and regulations covering the area. It can arrest individuals or ships of any nationality for defying those laws. This jurisdiction derives almost automatically from the jurisdiction exercised on the land itself. Without it, the law of the land would be difficult to apply: criminals could escape easily from arrest, customs and sanitation laws would be difficult to enforce, piracy against fishing and other vessels in the vicinity of the coast could not be prevented. Above all, it would be difficult or impossible to defend the land from attack from the sea. Partly for this reason the width of the area was for long defined in terms of the range of gunfire.

For years, the most widely accepted breadth of territorial waters was three miles, measured from the low water mark of the shore.

This was said to have been the range of artillery from the coast to sea in the eighteenth century. Some Scandinavian countries claimed four miles, and a few nations, especially in the Mediterranean, claimed six. However as ships became capable of greater speeds, as fishing industries demanded wider protection and as the range of guns increased, the width of the strip claimed began to grow. Even in tsarist days Russia had claimed twelve miles.

After the First World War, the extension of claims became more common. Some states asserted wider limits, not for general jurisdiction but for particular purposes; security, sanitary measures, or customs regulations. Others extended the territorial sea itself to twelve miles. An attempt to secure agreement on a common limit was made at a conference at the Hague in 1930, but this broke down without agreement. In general, the maritime countries, such as Britain, the United States and the Netherlands, with large shipping fleets operating all over the world, sought to preserve the traditional narrow limit of three miles to maintain the freedon of the seas. But governments with large fishing fleets operating in their own inland waters and those having special security fears tended to favour wider limits, in which they could protect their interests more easily.

After the Second World War, wider limits began to be claimed more frequently. Egypt, Ethiopia, Saudi Arabia, Libya, Venezuela and the Communist countries claimed a twelve-mile limit. Some wished to go even further. In the early fifties the countries of the west coast of Latin America became determined to preserve for themselves the highly productive fishing waters off their coasts. Because of the special character of the currents, these were uniquely productive but were increasingly threatened by the highly capitalized fishing fleets of the United States. El Salvador, in 1950, Chile, Peru and Ecuador in the Santiago Declaration of 1952, therefore proceeded to outdo all the rest by claiming limits of two hundred nautical miles for some or all purposes. Argentina made similar claims. In 1957 Indonesia claimed huge territorial waters in the areas between the many islands forming its territory, followed shortly after by the Philippines. By the time of the Conference on the Law of the Sea, which took place in 1958 (see below), thirty-five countries claimed limits of six miles or less, twelve claimed nine to twelve, and nine already claimed over twelve. After 1960, the widening of claims continued. A number of countries, especially in Latin America, extended their limits farther. New countries sometimes made wide claims: Guinea, for example, asserted a 130-mile limit. Few new countries claimed less than twelve miles. By 1967 more than half the sovereign coastal states claimed limits of twelve miles, with the rest fairly evenly divided between three and six. About half-a-dozen Latin American countries claimed two hundred miles for some or all purposes.[1]

Beyond this strip of territorial waters, a number of governments claimed, in addition, a *contiguous zone*, for customs, sanitary or 'security' purposes. Usually the two strips together gave a total of a twelve-mile 'maritime zone'. The contiguous zone was a relatively recent invention, resulting from unilateral claims to different areas for different purposes. Until 1958 there was no general international rule on the matter. But the 1958 Conference on the Law of the Sea passed a Convention on the Territorial Sea and the Contiguous Zone, laying down a limit of twelve miles for the latter and specifying the rights that a country could claim within it. A coastal state was given the right to exercise the 'control' necessary to protect 'infringement of its customs, fiscal, immigration or sanitary regulations'. This was the one clear-cut agreement on limits which that conference was able to reach.

In addition, a number of countries claimed special *fishing rights* in areas beyond the territorial sea. Within these, fishing was reserved to their own nationals, on an exclusive or preferential basis. This practice too became increasingly common from about 1950 onwards. The United States announced in 1945 that it reserved the right to declare 'conservation zones' beyond territorial waters, reserved for U.S. fishermen or for those who had fished there traditionally. Iceland in 1958 claimed a twelve-mile zone of exclusive fishing, which was for a time bitterly resisted by Britain, whose fishing boats had traditionally fished in these waters. Britain, however, in 1959 accepted the claim; and a few years later Britain itself claimed fishing rights up to twelve miles (though Britain made an exception for a number of European countries with which it entered into a special agreement). A number of other countries having three-mile territorial seas, including the United States, France, Denmark, Australia, Canada and others, claimed fishing limits of twelve miles.[2] Other countries claimed either

[1] For subsequent claims, see p. 147 below.

[2] The 1958 Convention on the Territorial Sea permits the establishment of a fishing zone up to 12 miles.

preferential or exclusive fishing zones, sometimes at much greater distances: for example, Korea claimed up to two hundred miles in certain areas. In other cases *conservation* zones were claimed, in which the coastal government reserved no exclusive rights for its own fishermen but reserved the right to restrict fishing to conserve stocks (India, Ceylon and Ghana claimed one hundred miles).

Finally, during the late sixties, one or two governments began to claim rights for 'environmental' purposes: that is to prevent pollution. Canada, for example, passed a law claiming such rights up to a distance of a hundred miles from its Arctic coasts.

The proliferation of conflicting claims stimulated various efforts to arrive at some international consensus on the different limits. From 1950, the International Law Commission (ILC) began to consider the question in detail and to draft international Conventions on the subject. This led in 1958 to the major conference on the Law of the Sea held at Geneva.

A limited measure of agreement was reached on certain subjects. Four Conventions were passed: on the High Seas, on the Territorial Sea and the Contiguous Zone, on Fishing and Conservation of the Living Resources of the High Seas, and on the Continental Shelf, but most of these were not very widely ratified. Even today none has been ratified by more than forty-seven states—barely more than a third of the world community.

The Conventions laid down the rights which governments could exercise within the various zones and proposed a system for deciding fishing disputes. But apart from the provision that the contiguous zone could not extend beyond twelve miles from the shore (more accurately, from the baselines from which the territorial sea was measured), no exact limits for these areas was attempted. On the most contentious and important point, the breadth of the territorial sea, no agreement was reached at all.

The three-mile countries, which were the largest single group, were insistent that other countries should accept their limit. When eventually compromises began to be proposed, none was able to secure the required two-thirds majority. Sweden proposed six miles, but the three-milers rejected this as too much. Canada proposed three miles of territorial sea, with exclusive fishing rights to twelve miles, which satisfied neither the arch-advocates of free seas, such as Britain, nor those which demanded twelve or more miles for security or other non-fishing purposes. Britain itself eventually agreed to accept six miles of territorial sea, provided full rights of passage for all vessels, including military, were assured between three and six; but this was unacceptable to twelve-milers, including Canada. Finally, the United States and Canada jointly proposed, as a last-minute compromise, a maximum of a six-mile territorial sea with a six-mile fishing zone beyond it. This was perhaps the offer most likely to succeed. But, since it was proposed that those who had historically fished in the six-to-twelve-mile area were to have rights there reserved to them, for an unlimited number of ships of unlimited size, this understandably did not appeal to the countries of Latin America, which felt threatened by U.S. or other fishing fleets.

A number of these proposals were voted on all in the same day, and amid considerable confusion and rancour. None got the necessary two-thirds majority. The Indians then proposed a general maximum of twelve miles. This is a suggestion which would now probably win overwhelming support, but at that time it was defeated. The Russians suggested that a state should choose its own limits which 'as a rule' should be from three to twelve miles; but this too was defeated on the grounds of its vagueness. So, though almost every possible alternative was proposed, none received the necessary support to pass a Convention. And an attempt to resolve the issues at a subsequent conference in 1960 also ended in failure.

There was little greater progress in relation to fishing rights. This too was a point passionately debated at the 1958 Conference. Because of the increasing resort to unilateral measures of 'conservation', the International Law Commission had produced a draft Convention on Fishing and Conservation of the Living Resources of the High Seas, which was discussed at the Conference. The traditional maritime countries and high-seas fishers, such as the United States, Russia, Britain, Norway and Japan, wanted the maximum freedom of fishing. The in-shore fishers and developing countries, especially the Latin Americans, resisted this. On some points a compromise was reached. Freedom of fishing in the high seas was reaffirmed, but subject to the 'interests and rights of coastal states'. A coastal state could take unilateral action to introduce conservation measures where only its own nationals were involved in fishing a stock. But if the nationals of other states were involved, it was to enter into negotiations with the governments of those states before introducing any such measure. If there was no agreement, there could be arbitration by a special commission, which would decide the question on scientific grounds related to conservation, and whose findings would be binding. If this did not take place, the government could introduce unilateral conservation measures, provided these were non-discriminatory. The compromise satisfied neither the distant-water fishers nor the coastal states. The Convention was not widely ratified: till now, only twenty-six states have done so. And the procedures it specified for resolving disputes were never once implemented even by those which did ratify.

Beyond the territorial sea, the contiguous zone, and any fishing zone that might be claimed, there existed, on the surface of the water, only *high seas*: that is, open water which not only could be freely navigated by all states (this was true even within territorial waters) but also was subject to no jurisdiction or control. The 1958 Convention on the High Seas laid down that 'the high seas being open to all nations, no State may validly purport to subject any part of them to its sovereignty'. Freedom of these high seas was said to comprise freedom of navigation, freedom of fishing, freedom to lay submarine cables and pipelines, and freedom to overfly for all states. But these freedoms were matched by certain obligations: for example, to ensure safety at sea, to assist those in distress, to avoid pollution, to prevent slave-running and to act 'with reasonable regard to the interests of other states in their exercise of the freedom of the high seas'. For even in the high seas, it was recognized, there was not total and absolute liberty for all to do as they pleased. This has important implications for the future.

Rights to the Sea-bed

So far we have been concerned entirely with the surface of the water. Let us look now, in slightly greater detail, at what international law says concerning the sea-bed beneath these waters.

Within territorial waters the rights of governments applied also to the sea-bed beneath them: that is, the government exercised full sovereignty in relation to the bottom of the sea as well as the top. The 1958 Convention on the Territorial Sea laid down that 'the sovereignty of a coastal state extends . . . to its bed and sub-soil'. But a government was under the obligation to grant some freedoms to others, even in these areas, most international lawyers contended: to lay down cables or pipelines, for example.

Until recently there was little interest in the sea-bed *beyond* territorial waters, since it could not be exploited, and international law had little to say on the subject. Soon after the Second World War, however, this situation radically changed. Immediately after the end of the war, on September 28, 1945, President Truman announced to the world that the United States was assuming rights of jurisdiction, for the purposes of 'exploration and exploitation', in the continental shelf off its shores. By this time, U.S. oil companies had already begun drilling in the comparatively shallow waters immediately adjacent to the coasts of Texas and California. They had become aware that these oilfields stretched far beyond the limit of territorial waters, and they had acquired the technical capability of exploiting them beyond those limits. It was important to the United States to ensure that foreign companies did not begin to exploit those areas, on the grounds that they were 'no man's land'—and so open to all.

The Truman Proclamation stated that 'the government of the U.S. regard the natural resources of the sub-soil and sea-bed of the continental shelf beneath the high seas but contiguous to the coasts of the U.S. as appertaining to the U.S., subject to its jurisdiction and control'. The Proclamation made a clear distinction between the surface of the waters and the shelf beneath it. It stated that no new rights were being claimed over the surface of the water; and that the freedom of the seas was in no way affected. On the sea-bed, however, the Proclamation declared that, to obtain effective exploitation of the resources concerned, it was necessary for the U.S. government to acquire jurisdiction. Therefore the United States would henceforth regard the resources of the shelf contiguous to the United States as though they were its own.

The Proclamation did not specify the outer boundary of the area claimed. As we saw, the term 'continental shelf' was originally a geological one and had been used for many years to describe a physical feature of the earth's surface. The assumption was generally made, based on an explanatory memorandum, that the Truman Proclamation referred only to this physical shelf. Beyond that area exploitation would in any case have been quite impossible at that time and seemed likely to remain so almost indefinitely.

Among other governments, the Truman Proclamation in general aroused little excitement or protest. This was perhaps partly because most were at that time distracted by other more pressing preoccupations. It was partly because the shelf remained largely unknown. And it was also partly because most governments—or all that were not land-locked—saw that they too might be able to acquire some of the same benefits as the United States by assuming rights of exploitation in their own continental shelf.[1] Indeed, most had perhaps a much greater interest in reserving such rights than the United States itself. For the United States, by virtue of its technological expertise in off-shore exploitation, was the one country which might have benefited from a totally free régime within the continental shelves all over the world. This would have given U.S. companies the chance to exploit sea-bed resources, including oil and gas, without payment of royalties throughout the world on the principle of first come first served. The continental shelves of the United States itself would of course have been exposed to competition from foreign oil companies; but these were technically so far behind that this loss would have been insignificant in proportion to the gains which the United States would have acquired elsewhere. In any case, a number of governments, especially Latin American, responded to the Truman Proclamation by asserting rights, usually in similar terms, but sometimes to areas far greater in extent than those claimed by the United States and sometimes, as we saw, including full sovereignty as well as exploitation rights.

Because of this widely shared interest, the principle of the right of exploitation for a coastal state within the continental shelf was relatively quickly accepted. By 1958, twenty countries had made claims to jurisdiction within their continental shelves. This was not, however, interpreted by most to include rights of full sovereignty. Many international lawyers held that, while it was reasonable to accord the coastal state the right to *explore and exploit* the resources of the shelf, this carried with it no rights of sovereignty, or general jurisdiction which might among other things have interfered with the traditional rights of states in other fields—in particular, freedom of navigation, fishing or scientific enquiry.

[1] The Proclamation sparked off some conflict within the U.S. between the Federal government and the states, especially Texas and California, which felt that they had an equal right to the resources of the continental shelf area. Subsequently, under Federal legislation, and legal actions brought by several states, the latter acquired rights up to a distance of 3 miles (3 maritime leagues in the case of Texas), while the Federal government exercised rights in the 'outer continental shelf' beyond. There have been similar Federal–Province conflicts in Canada and Australia.

Clearly, it was essential to establish a legal basis for the zone as soon as possible. Accordingly, in 1950 the General Assembly of the UN invited the International Law Commission to examine this problem, in conjunction with their study of the law of the sea generally.

The Commission considered the question over seven years. There were two principles that were suggested at that time for defining the outer limit of the shelf. One, fairly closely related to the geological concept of the shelf, was based on the depth of the sea. The other was based on the ability to exploit, and therefore considered the needs and capacities of the coastal states rather than the configuration of the sea-bed. The Commission considered both these possible definitions.[1]

A limit based on the depth of the water—for example a limit of 200 metres in depth or 100 fathoms—was at first suggested. This would have had the great advantage of providing a fairly precise and clear boundary which could be measured and known in advance. But the Commission felt there could well be further technical developments which would enable exploitation to take place beyond that limit. A reservoir of oil or gas might fall on both sides of the line, and it would be absurd to halt exploitation at a certain depth. It seemed more reasonable to reserve the possibility of according some rights beyond the line to the coastal state. It thus rejected a line based on depth alone, which it felt could destroy the 'stability of the régime'. It proceeded to choose an alternative even more unstable.

It decided to base its definition only on the capacity to exploit. It recognized that this must mean abandoning altogether the geological concept of the shelf. In its original draft, therefore, the ILC proposed that the area be defined as: 'the sea-bed and subsoil of the submarine area contiguous to the coast but outside the area of territorial waters, where the depth of the superjacent waters allows of the exploitation of the natural resources of the sea-bed and subsoil'.

This, of course, would give a perpetually changing limit, varying with every change in technology. Many governments, in commenting on the Commission's proposals, objected to the imprecision of this concept of 'exploitability'. Accordingly, the ILC in 1953 reversed itself, and proposed now instead a limit based on depth alone: the 200-metre isobath.

This, however, in turn aroused opposition among some governments, especially in North and South America. Several Latin American governments had by then asserted their claims to extensive limits, both on the surface of the waters and on the shelf, far beyond that distance. Moreover, on the west coast of Latin America the physical shelf is very narrow, so that a limit based on depth would have given them practically nothing. These countries therefore in 1955 called an Inter-American Conference on 'conservation of natural resources in the continental shelf and oceanic waters'. The conference reasserted the view, already expressed by some of them, that the coastal state should be accorded jurisdiction over a wide area, in the interests of conservation and the natural rights of states to neighbouring resources. This declaration apparently had a considerable influence on the Commission: it specifically noted the conclusions of the Inter-American Conference as one of the reasons for a further change of view.

In 1956, having tried in turn both the main possible variations, the ILC decided to combine the two. The definition it now proposed mentioned both depth and exploitability. That definition was finally adopted almost unchanged by the 1958 Conference on the Law of the Sea in the Convention on the Continental Shelf then adopted. In its final form this read as follows: 'For the purpose of these articles, the term "Continental Shelf" is used in referring (a) to the sea-bed and subsoil of the submarine area, adjacent to the coast but outside the area of the territorial sea, to a depth of 200 metres or beyond that limit to where the depth of the superjacent water admits of the exploitation of the natural resources of the said areas; (b) to the sea-bed and subsoil of similar submarine areas adjacent to the coasts of islands.' Within these areas the coastal state was to exercise sovereign rights 'for the purpose of exploring it and exploiting its natural resources'. This last phrase is important. The rights accorded were not full sovereignty but sovereign rights for a specific purpose: exploring and exploiting.[1]

The definition as it finally emerged, therefore, contained three

[1] The ILC, in proposing the text, said specifically that it had not felt able to accept full sovereignty for the coastal state over the sea-bed and subsoil of the continental shelf because of the possible effect of freedom of navigation and overflying rights. On the other hand, some individual governments (Argentine, Chile, and Peru), in extending their claims to the shelf, have purported to assert full sovereignty within that area.

separate defining elements: adjacency to the coast, the depth of the water and (alternatively) exploitability. But all this raised any number of difficulties. What was the relative importance of each of these? While the depth limit alone would have been relatively clear and specific, it was immediately contradicted by the exploitability limit. Given the exploitability limit, why mention depth at all? And once the depth limit had been abandoned, where did the exploitability limit end, if anywhere? If exploitation could be achieved at greater and greater depths, could national rights be extended outwards to the mid-point of the ocean or even beyond, to the point where they were being exploited on the other side? Some international lawyers began to argue that this was so. And if so, how would this be reconciled with the continued use of the term 'continental shelf' with its clear geological meaning, to describe what the Convention was supposed to be about? The ILC and the 1958 Conference quite deliberately retained this term against the arguments of those who sought to substitute the words 'submarine areas', which would have been more logical in the circumstances on these grounds.

Again, what was the significance of the words 'adjacent to the coast'? Did it mean simply 'opposite the coast', so prohibiting a nation from claiming or exploiting the shelf opposite other nations, without denoting distance; or did it mean 'close' to the land, in which case the right to exploit to the mid-point of the ocean would clearly be prevented? In that case, how close was adjacent? One hundred miles or one thousand? Did it exclude areas *under* 200 metres deep where the shelf stretched out for hundreds of miles, as it does in some places? In other words, how adjacent was adjacent?

Finally, what was meant by the words 'admits of the exploitation' of the resources of the area? Exploitation by whom? By the most advanced company or nation in the world, or only by the nation claiming the shelf? And what kind of exploitation is meant: the ability to extract any deposits of any kind by any means (core-drilling is now possible out to depths of 6,000 metres), or the ability to do so profitably? How was exploitability to be tested? If a government claimed an area on the grounds that it hoped to exploit it some time in the next century or two, did this represent a valid bar to another country exploiting the area, or must there be the prospect of *immediate* exploitation, and if so how immediate? If the latter was the case, how could technically backward countries hope to preserve their off-shore areas if the companies and governments of more advanced countries refused to assist them in immediate exploitation, and therefore claimed that the area was international and open to all? If not, how could any part of the sea-bed be kept from national claims if the coastal state claimed that advancing technology would enable it to exploit at some time in the future?[1]

The exploitability criterion was added partly because of political pressures, mainly from Latin American governments (most of which did not eventually ratify the Convention), and partly because it was recognized that exploitation beyond the 200-metre line would shortly be possible and should in some way be legalized. Because there was no conception at that time of the possibility of an *international* régime for the sea-bed (though one member of the Commission, Professor Scelle, favoured an international system), the only reasonable solution seemed to be, if exploitation was possible, to allow an extension of national rights.

The best way of looking at the 1958 Convention is perhaps to regard it as an interim measure, a first attempt to tackle the problem in the light of existing technical possibilities. It laid down a general rule of 200 metres in depth, but allowed some exceptions in cases where exploitation of a border deposit might otherwise be prevented, or exploitation might otherwise take place against the will of the coastal state. If exploitation was possible at all beyond 200 metres, it was probably still in the shallow physical shelf and should reasonably be reserved to the coastal state. This could at least prevent an attempt to exploit the

[1] Various ingenious ways of interpreting the Convention on the continental shelf were suggested to reduce the difficulties. It was argued, for instance, that the exploitability clause should be interpreted to apply to the physical shelf alone when this extended, as it sometimes does, beyond 200 metres in depth. But this ignores the fact that it was included partly as a sop to countries, such as those on the west coast of South America, which have no continental shelf. Others therefore argued that the shelf should end at 200 metres for those countries which have a geographical shelf, but allow the exploitability criterion as a possible alternative for those who have none or virtually none (for whom in fact exploitability would come much later). Another possibility suggested was to use a criterion of exploitability at a certain date, say 1958, when the Convention was passed, or 1964 when it came into effect: this at least would prevent the indefinite extension of the area claimed as shelf. All these interpretations suffer from the difficulty that there is nothing in the text of the Convention itself to justify them; nor would they remove the uncertainty in the present definition.

outer shelf of one country by the companies of another, perhaps more advanced. It could thus be regarded as the first stage in the effort to prevent the principle of first come first served or grab-as-grab-can being adopted in the sea-bed. If technical capabilities advanced still further, it would be necessary to draw up another instrument which would define the legal position in areas lying farther out. This is probably the most convincing (and the most historically accurate) interpretation of the Convention. It may be justified by laying the major stress on the word 'adjacent', and interpreting this in the sense of 'close'. This argument would support the view of those who hold that, with advancing technology, the law of the sea-bed must now be wholly remade.

But there were some authorities, on the contrary, who sought to regard the Convention as the final word on the subject. By *under*playing 'adjacent', and laying the main stress on exploitability, the continental shelf could be stretched, indefinitely, according to the level of technical capabilities. As one international lawyer put it: 'There is no room to discuss the outer limits of the continental shelf or any area beyond the continental shelf under the Geneva Convention. . . . All the submerged lands of the world are necessarily part of the continental shelf by the very definition of the Convention'.[1] Such views were increasingly quoted by representatives of large oil companies and others hoping to extend their own claims. In this view, the Convention has finally decided the legal problems not only of the shelf but of the whole ocean floor: the entire area with all its resources could be appropriated by the nearest nation, according to the principle of median lines. The effect of this would be to give huge sea-bed frontiers to countries with long coastlines, such as the United States, Brazil and India. It would give almost equally huge areas to small islands, such as Mauritius, St Helena and others, in the middle of vast oceans, which could have claimed rights over the sea-bed in all the surrounding sea, until they reached the area claimed by their nearest neighbours.

But not only is the meaning of the Convention disputed and obscure. Its value has been reduced for another reason. To this day it has not been widely ratified. In the first ten years after signature it was ratified by only thirty-seven states. This is only about one quarter of all sovereign states today. But it is not

[1] Professor Shigeru Oda of Tohoku University, quoted by Ambassador Pardo, in the UN General Assembly First Committee, November 1, 1967.

generally accepted, and could scarcely reasonably be maintained, that the Convention embodies customary law, and for that reason is binding even for those countries which have not ratified it. Many governments and international lawyers therefore, especially outside Europe and North America, deny the validity of the Convention as representing the state of international law on the question today. They claim that it is the unilateral invention of a few richer states which is not binding on others.

It was open to any Contracting Party to the Convention after 1969 to request a revision at any time under Article 13 (as it had then been in force for more than five years) and for the UN General Assembly to determine what steps should be taken in response to such a request. But the whole question of the limits of the continental shelf was by that time under discussion in the context of the sea-bed debate to be described shortly. And no attempt has in fact been made to secure a revision to clarify its meaning.

In many ways, therefore, the Convention had become a valueless document. Even if it had been generally ratified and acknowledged as binding, it was subject to so many and such conflicting interpretations that it could not have served as a useful guide to the action of governments. At the time when the sea-bed issue was first raised in the UN, the law defining the limits of national rights there remained both obscure and bitterly disputed.

Alternative Proposals for Defining the Continental Shelf

Because of the difficulties in any definition based on exploitability, a number of much simpler definitions were proposed for the continental shelf at various times.

Even at the Geneva Conference on the Law of the Sea in 1958 a number of delegations proposed a limit based on distance from the shore. This would not only be simple and clear-cut, but would also have the obvious advantage of securing some equitability among coastal states because, though it ignored the physical shelf, at least it would ensure that the areas acquired would be roughly propotional to coastline. Yugoslavia put forward a specific proposal for a draft article to limit the continental shelf in general to 100 miles beyond territorial waters. Norway and the UAR also wanted a limit based on distance. Both these were turned down. But some governments in recent times have revived the idea of a limit of this kind (see page 166 below).

Other countries at different times proposed limits based on depth alone without any mention of exploitability. This would of course give different coastal states quite different areas according to how steeply the shelf sloped. But it would have the advantage that the limit would continue to be based on the concept of the continuation of the land mass, and so would implement the original notion of a 'prolongation' of the land, which the International Court has since accepted. India, Norway and the United Kingdom all proposed at the 1958 Conference a limit of 550 metres (without any exploitability criterion). This was turned down by only 31 votes to 21. A number of other states, such as France, Italy, Canada and Argentina, were concerned that any definition should be clear and definite, and therefore also opposed the exploitability criterion. Yet other states (including Canada), which finally voted for that formula, did so only as a compromise because the definition they themselves favoured was unattainable. Conversely, a considerable proportion of those in favour of the definition finally adopted were Latin American states, most of which have not finally ratified the Convention. This raises in acute form the question whether nations should have the ability to influence the drafting of conventions unless they have the clear intention of ultimately ratifying it.

There was also a proposal at the Conference for a mixture of depth and distance. Yugoslavia at one time wanted a limit of 550 metres in depth or 100 miles in distance from the coast, whichever was nearer to the coast. This was overwhelmingly defeated at that time. But this again is the type of solution which has been revived more recently and may well finally win the greatest support. Many international lawyers have declared themselves in favour of a definition which combines depth and distance in this way.

Bilateral agreements among governments on the continental shelf have also done little to clarify the law on the subject. They have been almost entirely concerned with defining the lateral boundaries of the shelf rather than determining the seaward limit.[1] The seaward edge was of course not an issue between individual countries, and there was therefore no special reason

[1] They usually follow the principle which is laid down in Article 6 of the Convention that 'the boundary is the median line every part of which is equidistant from the nearest point of the baselines from which the breadth of the territorial sea of each state is measured'. This means that the choice and length

why agreements should have attempted to lay down the limits in this respect. In a number of cases they concerned areas, such as the Persian Gulf, all of which are below 200 metres and are therefore shelf under any likely definition. So long as the state of the law remains so disputed and uncertain, it is unlikely that any government will heedlessly prejudice future rights by specifying the exact seaward boundary of the shelf in such treaties.

So far as unilateral declarations of claims and municipal laws are concerned, a small number of states have confined claims to areas under 200 metres in depth—for example, Mexico and Portugal. Pakistan uses the limit of 100 fathoms. In some cases legislation refers simply to the 'continental shelf' without defining this. Other legislation, especially that passed since the signing of the convention, have borrowed its language: this is true, for example, of Italy, West Germany, Venezuela, Denmark and some of the Australian states. Many other countries, in their initial declarations and in subsequent legislation, provide no definition at all; this is true of the United Kingdom, the Netherlands and other countries. The U.S. government, which invented the concept of the shelf in its legal sense, is equally vague. Its Outer Continental Shelf Act of 1953 simply gave the Secretary of the Interior the power to grant leases for exploration purposes in the 'outer' continental shelf, without stating where this ends; and the Secretary of the Interior has in fact in some cases granted leases in areas in depths of over 1,000 metres. It is said that 28 countries have granted leases for oil exploration in depths of over 200 metres.

In other countries lip-service at least is paid to the idea of 'adjacency' contained in the Convention. Australian Federal legislation refers to the sea-bed and subsoil of the continental shelf 'contiguous to the coast'. Chile and Peru refer to the area 'adjacent to the coast' (even though the limits claimed are so enormous). Iran and Iraq both refer to areas 'contiguous' to the coast. But as no definitions of the terms are provided, they do not help much.

Thus a number of alternative definitions for the continental shelf area (and therefore, by implication, of the international sea-bed) have been proposed at different times. Some have been incorporated in national legislation claiming to define the shelves

---

of baselines can be of considerable significance in determining the respective shares of shelf resources. The 1958 Convention on the territorial sea lays down some principles governing the choice of baselines but leaves others undecided, including the maximum length of baseline.

of particular nations. More often, national claims have not defined the area in any precise terms, though national actions have sometimes implied a wide boundary of jurisdiction. But all the time exploitation in outer areas became more possible. And it therefore became more urgent that some international definition should be agreed which could restrain uninhibited unilateral claims and determine where national rights on the sea-bed end.

The Deep-Sea Bed

Beyond the continental shelf lies the deep-sea bed, which no nation can claim. This is the area described in recent UN resolutions as 'the sea-bed and the ocean floor and the subsoil thereof underlying the high seas, beyond the limits of national jurisdiction'. It is with the resources of this area that UN debates have been mainly concerned. What rights, if any, could national governments claim here?

The law relating to this area is even more unclear than that relating to the continental shelf, and for the same reason; it has not until recently been thought important. As one authority has said:

> As to the law governing the extraction of minerals from the deep-sea bed, the only certainty is that the law is uncertain. It is not clear whether any nation can acquire and claim sovereignty in the sea-bed, and if so, by what measures. It is not clear whether, without any claims of sovereignty any state (or private entrepreneurs) can lawfully proceed to dig and keep what it extracts.[1]

A distinction should be drawn first between the 'sea-bed and ocean floor' (there is no real difference between these two terms) and the subsoil. According to some legal authorities the sea-bed itself is subject to the same régime as the high seas; that is, it is generally free and open to all, and no action can be taken there which will restrict navigation. For example, Article 1 of the Geneva Convention of the High Seas of 1958 defines the high seas as 'all parts of the seas that are not included in the territorial sea or in the internal waters of a state'. The sea-bed can legitimately be regarded as a 'part of the sea' and therefore could be claimed as a part of the high seas. One of the main legal authorities on the law of the sea writes:

[1] L. Henkin, in E. Gullion, ed., *The Uses of the Seas*, *op. cit.*

A clear distinction must be drawn between the bed of the sea and the subsoil. As regards the former, the better opinion appears to be that it is incapable of occupation by any state and its legal status is the same as that of the waters of the open sea. The same reasons for maintaining it unappropriated in the interests of freedom of navigation apply with equal force to the bed of the sea.[1]

The only exception allowed, by this and other writers, to the rule of the non-appropriation of the sea-bed is in the case of a long-standing use of the sea-bed, for the purposes, for example, of pearl fishing.

During the discussion of freedom of the high seas in the fifties, it was frequently stated that among the relevant freedoms to be assured were freedom to explore and exploit there, and freedom to engage in scientific research. The Chairman of the ILC, during the discussions on this subject in 1955, said specifically that in the draft Convention on the high seas 'the Commission has merely specified the four main freedoms. It is aware that there are other freedoms, such as freedom to explore or exploit the subsoil of the high seas and freedom to engage in scientific research therein'.[2] This could be taken to imply that (unless some special régime is established) exploration and exploitation in the sea-bed are to be free and open to all on the basis of first come first served. Most others, however, maintain that because exploitation has not been possible until now, there is not yet any relevant international law relating to the sea-bed. And at least one other commentary by the Commission seems to give some substance to this claim. In its commentary on the same Convention, the Commission said: 'the Commisison has not made specific mention of the freedom to explore or exploit the subsoil of the high seas. It considered that, apart from the case of the exploitation or exploration of the soil or subsoil of the continental shelf, such exploitation had not yet assumed sufficient practical importance to justify such special regulation.' This implies not that there is freedom to exploit the deep-sea bed but that, because exploitation was still impossible there, it had not yet been necessary to regulate it.

[1] C. J. Colombos, *The Law of the Sea* (Longmans, London: 1967), p. 67.

[2] See UN Sect. document A/AC135/19/Add. 1, p. 12. The Chairman added: 'The Commission did not study this problem in detail at the present session.' As the UN report points out 'this freedom was not specifically included . . . as the subject was considered too premature and impractical for detailed regulation.'

If there were traditional freedoms of this kind, they would extend to the *subsoil* beneath the sea-bed. Some writers at least have suggested that national governments could 'occupy' part of the subsoil—for example, in laying pipelines or constructing tunnels. Colombos, for example, a leading authority, states:

> On the other hand, the subsoil under the bed of the sea may be considered capable of occupation. . . . It would therefore be unreasonable to withhold recognition of the right of a littoral state to drive mines or build tunnels in the subsoil, even when they extend considerably beyond the three-mile limit of territorial waters, provided they do not affect or endanger the surface of the sea.[1]

Article 7 of the Convention on the Continental Shelf quite explicitly provides that the right to tunnel beneath the sea-bed is not limited by any boundary of depth: 'the provisions of these articles shall not prejudice the right of the coastal state to exploit the subsoil by means of tunnelling irrespective of the depth of water above the subsoil.' But it is doubtful if this right could be used to justify exploitation by drilling of the subsoil in the deep-sea area beyond the shelf.

There is strong evidence that international law has never favoured the idea that coastal states could assert rights indefinitely in the area beyond the continental shelf. Even when the ILC was proposing to use exploitability as the *only* criterion governing the rights of the coastal state in the continental shelf in the early fifties, it never intended that these rights should extend indefinitely. At that time (1950) the Commission agreed on a concept of the shelf which made this clear:

> The Commission took the view that a littoral state could exercise control and jurisdiction over the sea-bed and subsoil of the submarine areas situated outside its territorial waters with a view to exploring and exploiting the actual resources there. The area over which such a right of control and jurisdiction might be exercised *should be limited*; but, where the depth of the waters permitted exploitation this should not necessarily depend on the existence of a continental shelf [my italics].

All the various drafts of the Convention and the commentaries by the Commission, though they differed in the basic principles applied, included the words 'contiguous to the coast', 'adjacent to the coast' or, in one case, 'in considerable proximity to the coast'.

[1] C. J. Colombos, *loc. cit.*

Moreover, even in submitting its final formulation (substantially the one eventually adopted), the Commission explicitly decided to retain the term 'continental shelf' 'because . . . it was not possible to disregard the geographical phenomenon, whatever the term—propinquity, contiguity, geographical continuity, appurtenance or identity—used to define the relationship between the submarine areas in question and the adjacent non-submerged land'.

This has now been reaffirmed. The decision of the International Court of Justice in 1969, determining the boundaries of various continental shelves in the North Sea, made clear that the limits of jurisdiction should be related to the physical shelf. In its judgement it said that 'what confers the *ipso jure* title which international law attributes to the coastal state in respect of its continental shelf is the fact that the submarine areas concerned may be deemed to be actually part of the country over which the coastal state already has dominion—in the sense that, although covered with water, they are a prolongation or continuation of that territory, an extension of it under the sea'. Although it cannot be said with any certainty, on the basis of this judgement, how far that prolongation can extend, it is at least clear that rights do not extend all across the sea-bed. There are, indeed, substantial arguments for stating that the only really pronounced and significant change of inclination in the sea-bed occurs at the edge of the geological shelf (usually between 130 and 200 metres depth); and it is, therefore, at this point that the 'prolongation' or 'continuation' of the land mass comes to an end. This geographical fact is of importance in arriving at a more satisfactory definition in the future.

Thus the position of international law on all maritime areas has remained shrouded in obscurity. The 1958 Convention on the Territorial Sea provided no definite limit to territorial waters. The Convention on the Continental Shelf was so ambiguous that it made the position on the shelf no clearer. On the deep sea there was virtually no law at all.

This was the confused and unhappy situation in which the law found itself in the autumn of 1967 when the sea-bed question was first raised as a major international issue.

CHAPTER 3

The Threat of Militarization

Sea-bed Missiles

There was a third problem, however, which the increasing capacity to occupy and use the sea-bed has made more and more urgent in recent years, and one in many ways still more alarming and more immediate than either of those we have been considering. This was the increasing possibility that parts of the sea-bed would be taken over and used for military purposes. The sea-bed could become the arena of a new arms race, as menacing and expensive as those which have afflicted other regions of the earth.

For some of the same advances in technology that made increasingly possible the exploitation of the economic resources of the sea-bed—the development of submarines and other vehicles which could explore and settle on it, the techniques which enabled men to spend considerable periods submerged beneath the sea, television and sensor devices which could give clear visual knowledge of the bottom of the sea, better methods of deep drilling and others—also created the possibility of using the sea-bed for military purposes. The capacity of one country to develop these techniques before another might, some believed, provide them with incalculable strategic advantages over its adversaries. And indeed a considerable proportion of the research on some of these techniques was carried out, at least in the United States and probably in the Soviet Union, not by oil enterprises and other civilians but by the armed services: this was itself an indication of the important military advantages which, it was believed, the military use of the region might afford.

Possible military uses of the area were of a number of kinds, some developments of traditional uses, some entirely new. The supreme advantage which the sea-bed could offer was concealment. At a time when missile technology had rendered every square foot of territory on land vulnerable to air attack, and when satellite reconnaissance made such surface areas subject to regular observation from the air, there was a special advantage in being able to conceal weapons, such as missiles, beneath the surface of the sea, where they were immune even from observation. As a UN Secretariat paper commissioned on the subject pointed out:

> The ocean is, for most practical purposes, opaque. Satellite surveillance has become a reasonably effective means for recurrent observation of land-based activities. It is comparatively less effective, however, for following or checking ocean sub-surface developments. . . At present, . . . there seems to exist a general appreciation that the concealment counter-measures are often more effective than the detection possibilities, and that this situation may prevail for some time to come.[1]

The most obvious use to which this concealment could be put was for the emplacement of strategic missiles. This was not, as has been sometimes suggested, a remote possibility in which the major powers were never seriously interested anyway. *The U.S. News and World Report* of October 16, 1967, reported that U.S. military planners were considering the possibility of sinking nuclear missiles in capsules under the sea 'off potential enemy coasts with a remote control mechanism for firing'. Another military expert in the same year forecast: 'Military installations are now centred reasonably close to the land mass: that will not be the case . . . ten years from now. We will carve out rather large chunks of the ocean away from the land masses which we . . . regard as very important to our national defence and . . . we shall deny . . . access by any other nation to the areas which we will block out'.[2] As one British military writer put it: 'It is possible that future developments in strategic defences will make new sea-based deterrent systems seem desirable . . . There are . . . proposals for unmanned missiles based on or under the sea. Thus the shrouding quality of the sea may very well be utilised in new

[1] Report of the UN Secretary-General, *Military Uses of the Sea-bed*, A/AC.135/28, pp 3–4.

[2]. Quoted in UN Assembly First Committee on Nov. 1, 1967 (A/CL/PV 1515).

ways.'[1] Numerous military writers referred to this possibility, in some cases recommending it as a logical next step.

Under-sea missiles might have clear advantages. As the UN paper put it:

> Strategically nuclear-weapon powers might, it is thought, find it desirable to replace land-based missile siloes with sea-bed bases or siloes. Such a shift might decrease the consequences to a nation and its population of a nuclear strike against its missile force. Also from the strategic point of view, a shift to missiles based on the continental shelf adjacent to a presumed adversary might make it possible to deploy shorter range missiles, perhaps a shift from ICBMSs [inter-continental ballistic missiles] to IRBMs [intermediate-range ballistic missiles]. This . . . might be regarded as advantageous from either a technical or economic point of view, or both.

Most such writings were concerned mainly with the offensive use of missiles: the use of the sea-bed as a new and more easily concealed emplacement for intercontinental rockets. Even if sited close to the coasts of the home state, they would have had this advantage. But additional benefits would be gained if the missiles could be sited, perhaps on seamounts beneath the surface, much closer to the territory of the potential enemy. This would provide greater accuracy and a shorter delivery time, and there would be much greater difficulty for an enemy in locating, homing on and shooting down the missile.

These advantages would be even greater in the case of defensive anti-ballistic missiles. For these too, if stationed halfway across the intervening ocean, would be able to shoot down the offensive missiles of the enemy at an earlier stage in their trajectory, perhaps before the independently targeted warheads in the MIRV—the multiple independently targeted re-entry vehicle—had detached themselves from the original missile. As one U.S. military writer put it: 'a near-bottom capability would permit deployment, covertly or overtly, at the option of the deployer and at the same time permit credible deployment within or out of range of the threatened protagonist in accordance with the level of political or military crisis.'[2]

Again, Polaris-type submarines might be semi-permanently

[1] L. W. Martin, *The Sea in Modern Strategy* (London: Chatto and Windus, 1967), pp. 33–4.

[2] J. P. Craven, 'Sea Power and the Sea-Bed', *U.S. Naval Institute Proceedings*, April, 1966, p. 49.

stationed on the sea-bed. As another U.S. strategic writer put it:

> Such developments may for example take the form of placing missiles as large or even larger than Polaris size on relatively shallow underwater barges on the continental shelf in such a way as to conceal their location. The underwater barges would move infrequently so that the potential of their being tracked by motion-generated noise would be minimized. Another possibility would seem to be a slightly mobile ocean-bottom system which creeps along. . . . Such systems can involve much larger missiles, might require underwater maintenance by personnel also located under water, might entail development of new kinds of detection and survival equipment to prevent attacks on emplacements, and so on.[1]

This would not be a very substantial change from the existing uses of nuclear submarines. But it would use the bottom of the sea on a semi-permanent basis and so would set a precedent for the military use of the sea-bed that could easily spread to other military fields.

A related possibility that has been discussed is the establishment of bases for submarines on the sea-bed. The UN Secretariat study quoted above reported that 'a logistic possibility which is sometimes discussed in the literature is that of sea-bed bases for nuclear submarines. Such bases would extend the length of time that a submarine may spend underwater and thus involve the possibility of concealed operation'. Here we see the logical conclusion of the use of sea-bottom vehicles. The sea-bed would now become subject to more permanent appropriation for the establishment of large-scale installations. The first step in the gradual occupation of the whole floor of the oceans would have been taken.

Conventional Uses

Though writings on this question are inevitably mainly concerned with the possible use of the area for nuclear weapons, the sea-bed could of course equally be used for missiles or submarines or other weapons bearing only conventional warheads. The same advantages of concealment and surprise could be won in this way. As the chief Soviet negotiator at the disarmament negotiations in Geneva pointed out: 'with the present rate of development of science and technology, one cannot rule out the possibility of the

[1] G. J. F. McDonald, 'An American Strategy for the Oceans', in E. Gullion, ed., *The Uses of the Seas*, *op. cit.*

emergence of new types of conventional weapons which could be used to strike from the sea-bed both at ships and at the territories of states'.[1] The U.S. delegate in the same discussions made another ominous forecast of future developments: 'Military and technical possibilities which may now seem remote may rather abruptly become imminent, and accordingly much more difficult to control.'[2]

One of the possible conventional uses of the area is for the emplacement of weapons aimed at foreign shipping or submarines. Manned torpedo batteries might be established on the sea-bed to threaten major sea lanes. These would enjoy the convenience and stability of a fixed base, and could make use of sonar and other detection devices, which could more easily be deployed from such stations than from within submarines. This could be especially attractive to a country such as the Soviet Union, which is particularly concerned with the capacity to wage warfare against surface ships.

Conversely, given the major threat of the Soviet submarine fleet to the western world, there could be a special attraction for western countries in the establishment of anti-submarine facilities among the seamounts and ranges at the bottom of the oceans. Anti-submarine weapons used in such bases could, in fact, be nuclear as well as conventional: many of those used today are already intended for nuclear warheads. As a British military writer has noted on this subject: 'Many of the ASW weapons are best suited for use with a nuclear warhead, and to some extent this reflects an effort to make up for inaccuracies in locating the target and homing upon it by employing a bigger destructive force.'[3] This anti-submarine role was another possible use of the sea-bed that began to be discussed. It became possible to imagine a situation in which larger and larger areas of the sea-bed were forcibly occupied by the forces of major powers for such purposes, even a competitive scramble to occupy the most strategically situated mounts or ranges.

Other possible uses of such manned stations that have been given by military spokesmen and writers are for 'submerged repair facilities, supply depots, re-arming and communications relay stations', or 'installation and operation of surveillance

[1] Eighteen-Nations Disarmament Conference, BV423, para. 26, July 29, 1969.
[2] *Ibid.*, para. 53, July 22, 1969.
[3] L. W. Martin, *op. cit.*

systems, both on the ocean floor and at Mid-depth'.[1] Many such projects for underwater naval stations are already under way. The Chrysler Research Department and the University of Miami are co-operating in a $100-million project called Project Atlantis, to build underwater stations for the U.S. Navy. The first station, 75 feet long, was to be placed in 1,000 feet of water off the coast of Florida in 1972 or 1973. A year or so later there were to be five of these stations at 6,000 feet, each housing ten to twelve men for 30 days.[2] They are designed to be command and control centres for anti-submarine warfare. Similarly, General Electric is engaged on a project called Bottom-Fix, designed to provide manned stations at depths of 3,500 metres, and eventually 7,000 metres. These are said to be required for a wide variety of deep-ocean missions, both military and scientific, and are intended to be operational by 1980. The U.S. Navy Civil Engineering Laboratory has been working since 1966 on a project for Manned Under-Water Stations (MUS), buoyant and anchored to the sea-bed and requiring a 30-ton 100-kilowatt nuclear power plant. There are also projects for undersea naval bases, built in tunnels extending out from on land and later extended upward to the sea-bed. Dr C. F. Austin of the U.S. Navy Weapons Centre declared that 'the Navy is thinking specifically in terms of sub-seafloor military bases operating surveillance gear, manning missile stations, and providing logistic support and staging areas for the undersea military forces of the future'.[3] Well over half the government-supported research on ocean science in the United States is today financed by the U.S. Navy. There is very little doubt that the Soviet Union is giving equally generous support to research on the military uses of the sea-bed.

One important use of the seamounts and other undersea areas, already fairly well-known, is for the emplacement of hydrophones and other devices for recording sound. The sounds are transmitted on shore and analysed by computers. It is known that the United States has set up installations of this kind close to the Soviet coast (for example, near Kamchatka) to record the passage of Soviet submarines from Soviet ports, and there seems every likelihood that the Soviet Union has done the same off U.S. waters. The United States has had a series of programmes for covering the seas around its coasts with such listening devices, beginning in the

[1] Robin Clarke, *Daily Telegraph*, 'The Deepest Secret', July 2, 1971.
[2] *Ibid.* [3] *Ibid.*, p. 15.

early sixties with Project Caesar and Project Colossus, followed by Sound System for Underwater Surveillance (SOSUS) and Project Trident. Similarly, there has been intensive study of noise emitters, which are used to distort and confuse the enemy's sonar devices. Even more complex undersea equipment is used to detect the anomalies in the magnetic field and the heat disturbances left behind by submarines. Attempts are also made to detect the presence of submarines by the use of echo sounders on the ocean floor: the U.S. Advanced Research Projects Agency has put down thirty seismic recording units on the ocean floor, ten of them in water deeper than 5,000 metres. All of these devices are frequently coupled to anti-submarine weapons, such as Subroc and Astor, both of which use nuclear charges. Finally, anti-submarine mines are also known to have been developed, which could make use of the sea-bed as well as of the waters of the sea. Given the huge importance of these various weapons in anti-submarine warfare (which could be one of the most significant forms of warfare in any major war of the future), it is not altogether impossible that there could develop physical battles for control of the seamounts, with each side seeking to destroy the instruments of the enemy, or to deprive him of use of such sites, and to defend and preserve their own instruments within such regions.

This possibility of a competition to acquire strategically advantageous mounts is not just fantasy. It has been seriously discussed and even proposed by military writers. One U.S. military author declared that within 'the ocean ridges and sea-mounts (6,000–8,000 feet) the capability to deploy vehicles and to mount installations is assured within the next decade'.[1] The UN study previously quoted, based on a wide range of military writings, reported that the need to ensure effective control of missile stations 'may motivate manned bases on the sea-bed, and . . . programmes are well underway, investigating the possibilities of prolonged stay, at continental shelf depths, of aquanauts'. Admiral Waters, an American, in commenting on U.S. experiments in keeping men under water for long periods, confidently predicted that by 1975: 'We will have colonies of [naval] aquanauts living and working . . . at depths in the neighbourhood of 1,500 feet, that is nearly 500 metres'.[2]

[1] J. P. Craven, *op. cit.*, p. 48.

[2] Quoted by Maltese Delegate in UN General Assembly First Committee, November 1, 1967.

Certainly, the effort and expenditure by navies devoted to experiments in enabling men to survive for long periods at considerable depths is difficult to explain except on the basis that some governments envisage the possibility of manned sea-bed establishments in the future. Experiments by the U.S. Navy during the 1960s, for example, brought big advances in such techniques. Men and animals have been trained to live in pressurized vehicles at very high pressures, breathing carefully controlled mixtures of helium, nitrogen and oxygen. In 1965–1966 the U.S. Navy undertook the experiments Sealab 1 and 2. In the first, a team lived at 200 feet under water for 9 days; and in the second, 3 teams spent 15 days each. Some men spent 30 days in pressurized conditions. Later experiments were designed 'to achieve, initially, an operational capability for men at 600 feet and eventually the capability to operate at 1,000'—in other words, beyond the continental shelf as normally defined. The British Navy undertook similar experiments, and there can be little doubt the Soviet Union has done the same,[1] though inevitably their activities are given less publicity.

The development of all these various techniques could thus lead to a new kind of military confrontation. Within a decade or so there could be permanent observation posts established on the sea-bed, emplacements for missiles, anti-submarine bases, and torpedo batteries for use against surface shipping, manned by teams which might stay for two weeks or a month at a time. A wholly new type of armed service might then come into existence. A new form of warfare would have been devised.

Extensions of Sovereignty

There has also been military research on new vehicles which could operate from such bases and could move around on the sea floor and at great depths. Wheeled or tracked vehicles, which could have important military uses, are known to have been developed. More than one military writer has commented on this possibility. The Secretariat study stated that

> It has . . . been pointed out that research and development is also underway with regard to vehicles designed to move around on the sea-

[1] Certainly, if they followed the view of the Soviet scientist quoted by the Bishop of Norwich in the House of Lords as declaring: 'the nation which first learns to live under seas will control the world'.

bed. Such applications . . . can be designed both for the purpose of nuclear weapons emplacement, and to carry out search and detection operations.

Vehicles for economic exploration are already in use. It is not altogether unrealistic to conceive eventually of some kind of underwater military vehicle coming into existence at some time in the future.[1]

On the other hand, there has been the development of submersibles, floating vessels that could operate at very great depths, at, or close to, the bottom of the sea. The U.S. Navy has given considerable support to research in developing these vessels. It developed the nuclear-powered, ocean-going submersible the *NR–1*, built for 'the maximum practical depth which can be assured within current technology', and which would be the 'pioneering prototype of the seabottom vehicles, which have all the requisites to revolutionise our concepts of the utilisation of the sea'.[2] The U.S. Navy also developed another type of vehicle, a welded pressure hull, of which the *Alvin* was an early example, capable of operating at 6,000 feet. Deep submergence rescue vehicles were developed, which did not require surface support. But perhaps most significant, the U.S. Navy began to develop 'a deep ocean 20,000 feet search submersible which would also be capable of mating with a conventional military submarine, providing a full ocean search and light work capability'.[3]

Finally, some previous military uses of the sea could be developed and given greater destructive power by the use of the sea floor. An important example concerns mines. Even in the Second World War use was made of ground mines; magnetic mines, which could explode without contact; and torpedo mines, which, when a ship passed over them, would surface and pursue the ship until they hit it. With the occupation of the sea-bed, it would be possible to establish mines fixed on the sea-bed, or even concealed beneath it, which could be released by the passage of a ship. If desired, some of these could be nuclear mines: the UN study referred to earlier states that 'another possible military use of the ocean floor indicated in the existing literature is concerned with nuclear mines, for offensive or defensive purposes. . . . Offensive

[1] The U.S. Navy has already developed 16 rescue vehicles for use on the sea-bed. These could well be adapted as vehicles for other purposes.

[2] J. P. Craven, *op. cit.*, p. 47.

[3] *Ibid.*, p. 47.

mines may be put in the coastal waters of another state. To avoid easy detection, such mines, it is thought, may be kept in readiness close to the bottom, rising to the optimum depth when activated . . .'[1]

Recognition of the possible military advantages to be obtained from occupation of the sea-bed has naturally encouraged discussion of greatly extended claims of sovereignty. The ambiguities in the 1958 Convention which allowed wide claims for economic reasons equally made possible extensive claims for military purposes. One American military writer concluded from the terms of the Convention that, given the rapidly developing underwater technology, 'it is clear that the ability to exert sovereign rights in the entire sea-bed has already received tacit approval'.[2] The same writer even held that this extension of sovereignty would affect the waters above the sea-bed: 'The right to regulate the jettison of material over occupied areas of the sea-bed, the right to regulate traffic above and around [such areas] and the right to discriminate between peaceful and belligerent traffic over occupied areas of the sea-bed may well result as emoluments of sovereignty'.[3] A British military writer similarly concluded 'that one result of such developments [the stationing of missiles on or under the sea] may be an effort to assert exclusive jurisdiction over areas of the sea or to establish identification zones analogous to those now maintained in the air'.[4] And the same writer commented later,

> It cannot yet be foreseen whether the technology in such fields as weaponry, rocketry, or underwater extraction of minerals may lead to further efforts to appropriate areas of the high seas for either permanent or temporary uses. Any such efforts would unquestionably extend the peacetime responsibilities of navies and generate requirements for new kinds of shipping, even if they did not lead to new occasions for international conflict. In view of the difficulty of underwater reconnaissance, the extensive policing of submarine areas would pose particularly challenging problems.[5]

Any such attempts to extend sovereignty, whether undertaken unilaterally by a particular nation, or by a group of more advanced

[1] UN Secretary-General, *Military Uses of the Sea-bed*, July 10, 1968, p. 5.
[2] J. P. Craven, *op. cit.*, p. 48.
[3] *Ibid.*, p. 49.
[4] L. W. Martin, *op. cit.*, p. 34.
[5] *Ibid.*, p. 138.

countries most likely to benefit, would obviously intensify the frictions and pressures which may anyway occur through the increasing militarization of the ocean. The growing possibilities of economic exploitation could also give rise to new conflicts and so possibly encourage further efforts at militarization, designed to assist the promotion of claims.[1] In other words, the very fact of the increasing use of the seas intensifies the possibilities of military conflict, and so inevitably increases the desire to extend military potential there.

Conclusion

There were good reasons, therefore, during the late sixties, for anticipating an arms race on the deep-sea bed: a race certainly involving many forms of conventional weapons (not always in conventional ways) and possibly the use of nuclear weapons as well.

This was a highly alarming prospect. On the one hand, the development involved an extension of the arms race into yet another area hitherto totally immune; on the other hand, it now affected an area previously regarded as belonging to nobody, and so, happily, beyond the range of all national claims or occupation. Quite apart from the purely military dangers which they bring, such military activities threaten marine life and peaceful fishing activities, as well as the economic exploitation of the area. Moreover, it is military activity in a form that is peculiarly obscure, hidden and mysterious, and which thus appears peculiarly sinister. It is, finally, a form of military power in which, from the nature of the techniques required, only the most developed countries of all can participate, and which therefore still further tips the military balance in favour of the most advanced and against the poorer and weaker states.

The longer such military activities continue, the greater the vested interest the nations concerned will acquire in them, and the more reluctant they will be to abandon them. Increasingly, it would be felt that such activities cannot be forgone without serious risk to security. Already in 1967 the U.S. Assistant Secretary to the Navy declared that the U.S. Navy had used the sea bottom 'for many purposes for many years', and that any agreement formally ensuring the peaceful use of the sea bottom could interfere with some national security enterprises.

[1] L. W. Martin, *op. cit.*, p. 175.

It was perhaps not surprising that this increasing military use of the sea-bed was a prospect that caused widespread alarm among many members of the international community. And it was one of the aspects of the whole problem which caused greatest concern among many member states when, in November, 1967, the question of the sea-bed was first raised within the UN.

CHAPTER 4

The Danger of Pollution

The Conservation of Living Resources

There is still another problem relating to the sea-bed and its environment and this relates to its conservation. There are two separate aspects of this problem. One is the conservation of fish and food—for example, from the dangers of over-fishing. The other is the conservation of the marine environment generally, from pollution of many kinds which increasingly threaten its natural ecology.

Both of these questions had begun to be discussed in UN bodies long before the problem of the ownership and control of the sea-bed was raised in 1967.

In conserving the *living* resources of the ocean the basic problem is that, where the economic motives of fishermen or nations are allowed free play without any kind of regulation, as has been normal until recently, some species will soon come to be threatened with total extinction. This has become an especially great danger in recent times with the development of large, highly mechanized fishing fleets, which, backed by ample capital, are engaged in very intensive fishing. Stocks decline to a low level and young fish may be taken, so threatening future stocks still further. Thus, measures to restrict or control the fishing of particular species have been discussed for some years. Some such measures have been implemented, but few of them have yet proved successful enough to remove the dangers to fish stocks.

A well-known example is the whale. This has been intensively hunted for about two centuries. Especially since 1945 it has been pursued avidly by highly capitalized and technically advanced whaling industries. In consequence the right whale and sperm whale are both almost extinct. The largest mammal of all, the blue whale, after intensive hunting in the last forty years, has almost suffered the same fate. The number caught declined steadily from 80,000 in 1930–1931 to under 400 in the mid-sixties. Thus a distinctive and unique species, the largest animal alive, has been almost totally eliminated from the face of the earth through the predatory commercial appetites of man.

Measures have been taken to tackle the problem for twenty-five years. They have not yet had much success, largely because of the fierce competition between the principal countries engaged in hunting and their reluctance to implement effective restrictions. In 1946, the International Whaling Commission was established, largely at U.S. instigation. Agreements were reached on close seasons and protected grounds, with annual quotas for each nation. Inspectors were carried on each ship. Quotas were based on 'blue whale units', one blue whale being reckoned as equal to two finbacks, or six other baleen whales. This was a system which may have seemed to give a fair balance of advantage, but reflected no known principle of conservation, as it could not determine the number of each species caught. It has now been amended. Moreover, in practice, because the Commission mainly represented the whaling nations, and because of the pressures of many of these, notably Japan, Norway and Russia, to maintain their catches, the quotas were kept excessively high. Sometimes the whaling fleets could not even catch enough adult whales to fill the quota. In consequence, every year many of the younger whales were killed and production therefore declined. At the Stockholm Conference in 1972, a resolution was passed calling for a total moratorium on whaling for 10 years, but this was not accepted by the whaling nations in the Commission. Although some new restrictions have now been imposed for particular species, the total whale population still seems likely to decline.

The salmon represents another well-known species whose survival has been threatened by uncontrolled sea fishing. Here the problem results from intensive fishing in the high seas, far away from the fishes' home spawning grounds. This often prevents adult salmon from returning to the rivers to spawn. Japanese fishermen in the Pacific, and Danish in the Atlantic, have been rapidly reducing the number of salmon returning to their home rivers in the United States and Britain. Attempts to reach an international agreement to restrict salmon fishing in the high seas, or the size of the nets used, have not yet been able to resolve the problem.

Over-fishing has also threatened a number of economically more important species in recent years. Cod in the North Atlantic have been threatened, and total stocks are likely to decline in the next few years unless there is a drastic reduction in fishing. Herring in the North Sea have been decimated through excessive fishing, and the industry virtually destroyed. For a time the halibut, which is intensively fished by the United States and Canada in the north-west Pacific, was threatened by a sharp decline in numbers, until both countries joined in forming the International Halibut Commission, which now operates a quota system. Here only two nations are involved (the Japanese have so far made no attempt to fish the halibut in that region), and it has proved easier than in other cases, for example whaling, to reach an agreement. There are also now fairly strict regulations governing the fishing of tuna, a large and economically valuable fish which has been harvested in great numbers in recent years, especially off both coasts of the Americas.

A belated awareness of the dangers of over-fishing led, over the years, to the establishment of a number of fisheries commissions to regulate activity. These commissions, mainly set up since 1945, have been formed either to regulate particular areas of the oceans, or particular species, or both. Among the regional bodies are the International North Pacific Fisheries Commission, the International Commission for the North West Atlantic Fisheries, the International Commission for the South East Atlantic Fisheries, the North East Atlantic Fisheries Commission, the Joint Commission for Black Sea Fisheries and others. Some regional bodies are bilateral: for example, the Japanese–Soviet Fisheries Commission for the North West Pacific; two between the United States and the Soviet Union covering the north-east part of the Pacific and the western areas of the middle Atlantic Ocean; and three interlocking agreements between the United States and the Soviet Union, the United States and Japan, and Japan and the Soviet Union concerning the king crab. In general, there is now a move towards confronting the problems of individual species, each involving special problems, rather than those of particular areas. Thus in recent years, commissions have been established, for example, for the inter-American tropical tuna, the Pacific tuna, the Atlantic tuna, the international Pacific halibut, and the international Pacific salmon, the North Pacific fur seal and so on.[1]

Such bodies are set up under international conventions, and the powers they wield vary from one to another. Most of the commissions have the power to prescribe closed areas or close seasons, to regulate the type of fishing gear used, to lay down the minimum size of fish that may be caught or the size of mesh of fishing nets and trawls. Without regulations of this kind, young fish may not survive, and future generations of fish will be threatened. In some cases (for example, salmon fishing in the North Pacific) there is an attempt to control the total volume of the catch, sometimes by sharing it out in agreed proportions among the two countries mainly involved (in this case the United States and Canada). The Soviet Union and Japan share out the catch of crabs in the north-west Pacific. The International Whaling Commission negotiates the proportion of the total catch allowed to individual countries. In the case of seals in the North Pacific, all sealing in the oceans is now forbidden, and the two countries which catch seals on land, the United States and the Soviet Union, deliver 15 per cent of their gross take each year to Japan and Canada, which now catch none at all. In the north-west Atlantic a system of quotas for individual countries to restrict the catch has now been introduced. This has required amendments to the Convention governing the Commission involved to increase its powers.

Sometimes, especially in the Pacific, the Commissions have established their own enforcement agencies to ensure that the regulations are applied by the fishing fleets of individual countries. This is usually done by a system of mutual inspections. Authorized officers of one contracting party, on board vessels flying its flag, may board and inspect the vessels of another party. Sometimes such officers, if they are satisfied than an infringement of the conservation measure has been committed, can seize the vessel or arrest those on board and hand them over to their own govern-

[1] Since 1960, some commissions have been established directly under the auspices of the FAO: for example, the Indo-Pacific Fisheries Council, the General Fisheries Council for the Mediterranean, the Regional Fisheries Advisory Commission for the South West Atlantic, the Indian Ocean Fishery Commission, the Fishery Commission for the Eastern Central Atlantic and so on.

ment, which has the responsibility for enforcing the agreement.

Over and above the Commissions there stands the FAO. Within the UN family, this undertakes responsibility for fisheries, including fish conservation. In 1966, the FAO established its own Committee on Fisheries. This has reviewed arrangements for co-operation in the exploitation and conservation of fish, both between individual governments and among the various commissions and inter-governmental bodies, especially in management, development, research and statistics. It has come to the general conclusion that conservation and management should be based initially on individual stocks[1] though both the Committee and the FAO as a whole believe that there is no universally applicable system for managing fishery resources. The FAO also seeks, through its Sub-Committee on the Development of Co-operation with International Organizations Concerned with Fisheries (that is the Fisheries Commission), to co-ordinate its own conservation work with that of the Commissions to improve contacts among the Secretariats and the flow of information on such questions. The FAO has now decided to try, as a first step, to improve generally quantitative knowledge about existing food stocks in the sea, and the potential harvest for the future. Advice on these questions is provided by a non-governmental body, the FAO's Advisory Committee on Marine Resources Research (ACMRR), a body of fifteen experts appointed in their individual capacity. The International Council for the Exploration of the Sea, which is primarily a scientific body and is outside the UN, is also concerned with determining fish stocks.

Finally, the UN itself has shown some concern on the general question of the conservation of fish. In 1958 it passed a resolution on the resources of the sea, which called for intensified action in conserving stocks. But the natural desire of developing countries to expand their fishing fleets and take a larger share in world fishing, together with the reluctance of the richer countries to reduce their own share, as well as ever-advancing technology, have the effect that over-fishing is increasing all the time. As one recent report described:

> Ocean fisheries continue to expand with frightening rapidity. . . . Ocean fishing power on a world-wide basis is growing at a much more

[1] United Nations, *Exploitation and Conservation of Living Marine Resources*, E/48482, May 12, 1970.

rapid rate than the means of measuring its effect on the fish stock it is being applied against. . . . This whole field of marine science is being swamped by the developing fishing power. The nations devote their ocean research funds to the development of fisheries, but they are laggardly in providing research funds for the detailed biological and population dynamics research which alone can give guidance in the solution of the problems which expanding fishing effort creates. . . . Nations do not like to put their fishermen under regulation, even to provide the conservation that they have agreed to provide, unless the scientific need for the regulation is established, and they do not like to put up the money to provide the research needed either to determine the need for regulation or the form it should take.[1]

Despite the undoubted truth of these remarks, it is scarcely surprising that new countries especially, which find themselves far behind richer countries, both in the size and sophistication of their fishing fleets, are reluctant to see regulations imposed which may restrict their entry into this field, just at the moment when they are beginning to expand their capabilities.

Because of the increasing depletion of some important stocks, a number of experts have begun to suggest that chronic over-fishing in the sea will be prevented only by the establishment of a more effective and powerful international authority to manage the world's fishing stock. One book, by two experts on fisheries, set out a detailed plan for an authority of this kind.[2] This suggested that certain countries might be given a monopoly in the fishing of certain zones, for certain species, on behalf of mankind, and the profits distributed on a mutually agreed world-wide basis, somewhat comparable to the system already adopted for fur seals described above. This would mean that the body or industry responsible for each species could concentrate on the most efficient means of exploiting that particular species without fear of excessive competition, but would have a direct interest in ensuring effective conservation measures against over-fishing. At present, competition gives an incentive to fish as much as possible without regard to the future, since conservation cannot be con-

[1] FAO, Report on 'The State of Ocean Use and Management', by W. M. Chapman, 1967.

[2] F. T. Christy and A. Scott, *The Common Wealth in Ocean Fisheries* (Baltimore, Md.: Johns Hopkins, 1965). See also J. Bardach, *The Harvest of the Sea* (London, 1967), p. 232: 'Perhaps the most appealing solution from the point of view of both economics and conservation would be a truly international authority for harvesting and managing the living resources of the sea.'

trolled by one nation or fleet and each industry wishes to make sure that it gets the maximum share of what is available.

The Conservation of the Marine Environment

The problem of conservation of the marine environment has only more recently come to be recognized. Let us consider first the nature of the problem, and secondly look at the various ways in which it is being tackled.

The type of pollution of the sea which has received perhaps the most publicity is that by oil, from tankers and other ships. This can arise from two main causes: either from accidents, such as the *Torrey Canyon* disaster, in which an oil tanker is in collision or runs aground; or, more commonly, because of the deliberate discharge of oil from tankers or other ships on the high seas, especially through the washing out of tanks. Deliberate discharge of this latter kind may come to decline: an increasing proportion of new tankers today use the system known as 'load on top', which makes it possible for tanks to be washed out without discharging oil into the sea. New international measures have also been taken or are planned to control the problem. But a considerable amount of deliberate discharge nonetheless still takes place at present. Against this, pollution through accidents is almost certain to increase, because the volume of oil shipped by sea is continually growing. Of the rapidly increasing production of oil today, over 60 per cent is transported by tanker; and as the average size of tanker increases all the time (from under 20,000 tons in 1950 to 500,000 today, and possibly a million shortly), the average size of oil spills through accidents must also increase. One estimate has put the total of oil that enters the sea through accidental or deliberate discharge at a million tons a year.[1] Other estimates range between 1½ million and 3 million tons. The total amount of oil entering the sea from all sources, including those on land, is put in some quarters at 10 million tons.[2]

Related to this danger is the growing threat of oil pollution through oil drilling and production on the sea-bed.[3] Serious

[1] *Marine Pollution*, Note by the Secretary-General E/37/2, All 7 of Jan. 11, 1971.

[2] M. Blumer, 'Oil Pollution of the Oceans', *XV Oceans* (Woods Hole Oceanographic Institution), October, 1969.

[3] There is also some escape of oil in the oceans through natural seepage of oil through the sea-bed, as in the Santa Barbara Channel, for example, and through the decay of marine plants and animal life.

incidents through blow-outs of wells at sea have so far been comparatively rare, but they are likely to grow in number with the rapid increase in off-shore production, which we examined in Chapter 1. Between 1954 and 1969, about 25 blow-outs occurred among the 8,000 wells sunk in off-shore areas of the United States, but serious pollution occurred only in two cases. One of these, however, off Santa Barbara, resulted in the contamination of an area of nearly 1,000 square miles through 230,000 gallons of oil. Similarly the increasing use of oil pipelines and oil storage tanks beneath the sea may in time lead to discharges of a very serious kind through damage to the pipeline, whether accidentally or through sabotage, close to a coast. As production extends out into the international sea-bed, this danger will clearly grow.

The damage that oil pollution causes is of a number of kinds. The most obvious is the contamination of coasts and beaches, when the oil is washed on shore, which is not easy to treat and can cause considerable economic damage to coastal resorts. But there are many others, having more long-term effects. Sea birds are very badly injured by oil, so that, even when extensively treated and cleaned, they rarely survive for long. Although in general fish are little affected by crude oil, they are killed by fuel oil, and through chronic exposure to oil may be damaged or made so foul-smelling as to be unfit for consumption. This in turn can affect other forms of life. Chronic oil pollution, which occurs in harbours, and other areas where many ships pass, can badly affect all forms of life, by removing oxygen from the water. Another serious effect of oil pollution, where the oil does not reach the land, and is not oxidized, is when it forms large tarry lumps in the water which remain for several months. Many sea travellers in recent years, including Sir Francis Chichester, and Thor Heyerdahl on the *Ra*, have reported how widespread is the pollution of water with 'tarlike or asphalt-like' material.[1] Finally, the marine ecology generally can be seriously disturbed by the large-scale presence of oil. As one report said: 'There is some indication that chronic pollution from such forces may produce irreversible effects, and pending results of further inquiry, it seems desirable to prevent even slight escapes of oil into the sea.'[2] At present these long-term

[1] O. Schachter and D. Serwar, 'Marine Pollution, Problems and Remedies', *American Journal of International Law*, January, 1971.

[2] Report of the Secretary-General, *Marine Pollution*, A/7924 of June 11, 1970, p. 4.

effects are not at all well understood, especially those from the oxidation of oil, which may deplete the oxygen supply on which much marine life depends.

More dangerous to human health than oil pollution is that which occurs through the emptying of sewage and industrial wastes into the sea. Because available areas for dumping sewage and other wastes on land are becoming depleted, and because of the growth of population in large cities close to the coasts, the volume of material dumped in this way is increasing rapidly. Fortunately, the salt sea water seems to have some purifying effect on sewage, at least in moderate climates, and direct danger to health is less than is usually supposed. But where the sewage is inadequately treated, it can lead to bacterial and virus infections, as well as to the degrading of beaches and of the sea waters themselves. Even when the sewage has been treated, it can still cause infection indirectly, especially in tropical areas—for example, by its effect on shellfish harvested in the contaminated area. The bacterial count in New York Harbour has increased ten times between 1948 and 1968, probably because of nitrogen from the land run-off; four million tons of sewage sludge are now pumped into New York Harbour every year. Worse, the discharge of sewage leads to an increase in the total quantity of nutrients in the sea water and so to the volume of plankton. This in turn can lead to a poisoning of shellfish and other secondary species, or disturb the entire ecology of the region by the artificial increase of particular species: huge rashes of foul-smelling red weeds have occurred in some areas—for example, off Florida—because of artificial fertilization of this sort, and this drives away other forms of life, including fish.

Industrial waste, whether discharged direct into the sea or from rivers into the sea, can be even more serious. The effect can already be seen at its worst in the Great Lakes of North America, the Baltic and Black seas, all enclosed seas close to large industrial regions, which have by now become virtual wastes in which oxygen and life have been almost totally eliminated. National legislation controlling or prohibiting the disposal of industrial wastes on land or in rivers often only encourages more dumping at sea. Very many materials are discharged direct into the oceans or into rivers that feed into the sea, and pollute both equally. Some dangerous substances, such as chlorinated hydrocarbons, pesticides and cyanide are still often deliberately taken out on boats and dumped. Mercury, and to a lesser extent copper and zinc, can be accumulated by fish in huge concentrations, and if the fish are subsequently eaten by humans, can cause injury or even death: very serious cases of mercury poisoning, including 50 deaths, have been caused in Japan in this way. Carcinogenic substances derived from the wastes of the petro-chemical industry are transferred to fish and other organisms and, when later consumed by humans, cause danger to life. Nine species of fish have been made unfit for human consumption by mercury and many Japanese are now reluctant to eat fish.

The marine ecology as a whole can itself be badly disturbed. For instance, wastes containing sulphur can altogether destroy shellfish. The migration of salmon and other fish is disturbed and distorted by industrial waste. Pollutants may absorb oxygen, thus suffocating living creatures that depend on it. Others simply poison animals or plants. With growing industrialization all over the world, and the increasing concentration of populations near the coast, all of these problems are increasing rapidly. In the United States alone, 'the volume of industrial wastes already twice the volume of domestic wastes, are expected to increase 7-fold within a decade'.[1]

Pesticides which are either blown by the wind from the land to sea, or run off into rivers and so to oceans, are equally damaging. These are used in rapidly increasing quantities. DDT, dieldrin, endrin, and other chlorinated hydrocarbons can become highly concentrated in plants and animals because they are not metabolized. Even if not lethal, they can affect the reproductive process, and so disrupt a whole food chain. A number of species of birds are already threatened from this cause. Similarly, the discharge of detergents which have a high content of phosphates, and fertilizers containing nitrates, can over-fertilize and so affect the balance of nature within rivers and seas, creating serious pollution as a result. The so-called red bloom, which has occurred increasingly off the Gulf Coast of Florida and in other places in recent years, is sometimes caused by the blooming of phyto-plankton fertilized in this way. This can cause irritating smells and the destruction of other species. Such artificially stimulated plants may also, when they finally break down, use up oxygen in the water, and so defertilize particular areas: this is one of the major causes of the destruction of marine life in the Baltic.

[1] O. Schachter and D. Serwar, *op. cit.*, p. 27.

A particularly grave danger is concerned with radio-active wastes. These are customarily dumped at the bottom of the oceans in lead containers wrapped in cement. They could, of course, be lethal if any of the radio-activity escaped from the containers in which they are dumped; at least one such container has been washed ashore, and another brought to the surface by a fisherman. In addition, the effluents from atomic reactors sometimes contain low levels of radio-activity. In a few cases the International Atomic Energy Authority has reported a high level of radio-active caesium from such causes. In a more general way there is the danger of radio-activity occurring from the use of radio-isotopes, nuclear explosions and other causes, becoming concentrated in fish or other marine life. The huge increase in the use of nuclear power (expected to rise by 1,000 per cent this decade) must multiply the risks. Although the United States has now limited the dumping of radio-active wastes in the oceans, it continues to be undertaken by European countries. And the United States has itself recently dumped highly toxic nerve gases in the Atlantic. In addition, there are large amounts of other chemicals which are dumped by commercial firms in containers in the ocean. It seems not unlikely that eventually a case will arise where the container is not secured and the chemicals escape.

Another new source of contamination is the modern means of surveying the ocean floor. As we saw in Chapter 1, a common method of surveying is through seismic explosions. All exploration of oil, gas and other resources is preceded by a seismic survey of this sort. This can lead to the destruction of plants and fish. New devices are now being developed which have less effect on the surroundings, but the greater volume of surveying now undertaken could nevertheless lead to increasingly serious damage to the ecology of the sea floor. Similarly, 'thermal pollution' from the heating of water by power stations causes a decrease in aeration and can encourage blooms of poisonous plankton.

Some of the agitation expressed in recent times about threats to the environment may well be unduly alarmist. Much of it underestimates the action already being taken to meet such dangers. But there are a number of reasons why, though marine pollution may have damaged few human lives so far, it could have serious effects unless vigorous action is taken soon. The volume of materials entering the ocean is today increasing at enormously high rates, partly because of the exhaustion of sites for dumping on land, and increased controls on dumping there, but mainly because of rapid industrialization. The increase in production of petro-chemicals, plastics and fertilizers is especially great. If the volume of total industrial wastes from the United States is to increase sevenfold this decade, the amount reaching the sea may increase as much or more. As we saw, pollutants sometimes become very highly concentrated in animal organisms within the sea. These seem able to absorb high quantities of particular substances through a process known as bio-accumulation. The concentration of materials which comes about in this way varies from a few hundred to several hundred times the concentration in the surrounding environment. At the same time, pollutants can be quickly carried and dispersed over wide areas. Fish and penguins have been found to contain extraordinarily high concentrations of insecticides and other pollutants which must have originated thousands of miles away.[1] Pesticides applied on the African continent have been detected off the coast of India and in the Caribbean, having been carried by the prevailing winds. Deep currents in the ocean can carry wastes from mining and dredging activities over large areas.

Thus the problem of ocean pollution is becoming more serious all the time. Because the seas are the world's ultimate sewer, much of its muck and poison finally ends up there. Once exploitation of the sea-bed begins to take place on a large scale, the problems will be far worse. The dangers of leaks or blow-outs during oil and gas drilling, of metal contamination in the mining of copper and mercury, will obviously become many times greater than today.

International Organizations and Pollution

When the problem of the resources of the deep sea was first raised at the UN in 1967, international organizations had for many years been concerned with various aspects of pollution. But this activity had been scattered and divided among a number of different bodies.

Since its foundation in 1959, the Intergovernmental Maritime Consultative Organization (IMCO) had been dealing with the problem of pollution of the sea from ships, especially from oil. The first international agreement dealing with this had been drawn up in the 1930s. A more important, and more widely

[1] J. Bardach, *The Harvest of the Sea* (London: Allen & Unwin, 1967), p. 246.

ratified, Convention was agreed in 1954. IMCO was the body responsible for seeking to apply the Convention, which was amended in 1962 and in 1969, to deal with deliberate discharge of oil. Until recently, the Convention prohibited discharge of oil and oily mixtures from ships except in designated free zones. With the amendments now passed (in 1969), it prohibits all discharge of oily wastes on the high seas, except under a few specified conditions. It also requires special fittings to prevent the escape of oil, detailed records of the treatment of oil, and the right of inspection. These latest amendments have not, however, so far been widely ratified, and have still not come into effect even for those nations which have ratified them.

To cover the discharge of oil through accidents, two new Conventions were drawn up after the *Torrey Canyon* disaster in 1967. One gave a country the right to take action on the high seas to protect its coast against damage from pollution. The other established the liability of owners to pay compensation in the case of oil spills through accidents, up to a limit specified, to be covered by insurance for the purpose. At the same time, a new compensation fund was set up to which ship-owners would contribute, and which would ensure that adequate compensation is available where an accident occurs. (In the case of the *Torrey Canyon*, Britain and France had to negotiate for an out-of-court settlement from the shipping company concerned.) In addition, IMCO produced a two-volume report on measures to combat pollution at sea. It examined new methods of oil storage in ships, and ways to detect and punish deliberate emissions of oil in the high seas. It inaugurated research on methods of removing oil from the sea and protecting coasts when a spillage had occurred. It proposed traffic separation schemes, for example in the Straits of Dover and other congested areas, to reduce the risk of accidents. And it considered problems relating to the construction and operation of drilling rigs to reduce the risk of blow-outs.

The World Health Organization (WHO) was responsible for the question of atmospheric pollution, including that affecting the seas by way of the atmosphere. It has had a committee that examined air pollution, and a centre in London to undertake research and collect information and statistics. It also gave assistance and training to developing countries on sewage and waste disposal in coastal areas so as to minimize pollution of the sea.

The International Atomic Energy Authority (IAEA) was concerned with pollution from radio-activity in the oceans and the means of controlling the dumping of radio-active wastes. It attempted to keep a register of releases of radio-active solids or liquids into surface waters. It established guides and standards concerning the methods used in transporting and dumping such wastes. A conference and two panels of experts proposed a system for monitoring the disposal of radio-active materials in the ocean. However, despite a number of studies designed to lay down standards for the dumping of radio-active wastes, and strong recommendation of more than one committee on the question, the IAEA consistently declined to establish any such general rules.[1]

The World Meteorological Organization (WMO) was concerned with those aspects of the ocean which affect the weather. It has established weather stations in different parts of the oceans, fitted with instruments to give readings on atmospheric pressure, temperature, humidity, and other measurements of importance in weather forecasting. Because of the use of these stations in providing information of other kinds about the ocean, the WMO has had close contact with other international bodies whose concern is with oceanology, the science of the oceans. For example, it co-operated with UNESCO in developing an Integrated Global Ocean Station System (IGOSS): a system for monitoring the ocean through automated buoys, equipped with detectors for gathering information and for transmitting it throughout the world.

The main body responsible for the scientific investigation of the oceans was the Intergovernmental Oceanographic Commission (IOC). This was established only in 1960 and, like some other scientific bodies, was placed under the general supervision of UNESCO. Its purpose is 'to promote scientific investigation with a view to learning more about the nature and resources of the oceans, through the concerted action of its members'. In other words, it tries to encourage its members to co-ordinate and arrange their own programmes of research in such a way that they fit in

[1] Countries which have ratified the Convention on the High Seas are under an obligation to prevent the pollution of the high seas by radio-activity, taking into account any standards and regulations that may be formulated by the competent international organization. Since no standards have been laid down by IAEA, it is doubtful if this represents a very meaningful obligation.

with one another rather than undertaking its own activities. It has co-ordinated a number of expeditions, including the Indian Ocean Expedition and a Co-operative Investigation of the Tropical Atlantic. It now has over 70 members. Though it remains under the general auspices of UNESCO, the other specialized agencies which are concerned with affairs of the sea have recently been given increasing influences in its operations. It has its own international co-ordination group on pollution, Global Investigation of Pollution in the Marine Environment (GIPME), which seeks to examine the causes of pollution and the way it interferes with the natural ecology of the ocean.

UNESCO has, in addition, its own Office of Oceanology. This undertakes a number of programmes in the field, and provides technological assistance to developing countries, mainly in the form of consultants and experts, or of fellowships for representatives of these countries in developed countries.

Another scientific body set up more recently is the Group of Experts on the Scientific Aspects of Marine Pollution (GESAMP). The main organizations represented on this are IOC, IMCO, FAO, UNESCO and WMO, all of which are concerned with pollution of the seas in one way or another. As its name implies, its concern is primarily with the scientific aspects of pollution, the study of the processes by which it takes place and the nature of the polluting agents.

There are also a whole range of non-governmental scientific bodies with an interest in this field. In 1957, the International Union of Geodesy and Geophysics organized the International Geophysical Year, which included studies of the ocean. This led to the formation of the Special Committee on Ocean Research (SCOR), a more permanent body of scientists with special interests in this field, operating under the general auspices of the International Council of Scientific Unions (ICSU). SCOR organized the Indian Ocean Expedition (1959–1965) until this was later taken over by the IOC. Moreover, there are a number of more specialized bodies, International Council for the Exploration of the Sea, the International Association of Biological Oceanology (IABO), the International Association of the Physical Sciences of the Ocean (IAPSO), the International Hydrographic Association, based in Monaco, and concerned with mapping the oceans, and the Commission on Marine Geology, among others.

Finally, the UN itself has a continuing interest in the oceans.

The series of resolutions passed since 1958 on the resources of the sea called for measures to tackle the pollution problem and the integration of international efforts in this field. In addition to this a number of international conventions concerned with pollution were drawn up by UN conferences. For example, Article 5 of the 1958 Convention on the Continental Shelf declared that 'the coastal State is obliged to undertake, in the safety zone, all appropriate measures for the protection of the living resources of the sea from harmful agents'. Article 24 of the Convention on the High Seas laid down that 'every State shall draw up regulations to prevent pollution of the seas by the discharge of oil from ships or pipelines, or resulting from the exploitation or exploration of the sea-bed and subsoil, taking account of existing treaty provisions on the subject'.

Thus international bodies were acquiring growing responsibilities in this field. National governments were sometimes beginning to adopt more effective national measures to confront the problem. But there was little effective co-ordination among their various activities.

The Main Difficulties

By the mid-sixties, therefore, there was already a widespread awareness that there existed a serious and growing problem of marine pollution. There was the beginning of an international institutional structure for dealing with it, although this remained confused and unco-ordinated.

There were many factors contributing to the difficulties. There was a general deficiency in the basic knowledge required to cope with the problem. This applied even to the sources of some pollution, and of the ways in which it could be counteracted. There was considerable ignorance about the movement of waters and currents within the oceans, and of how various pollutants were transmitted great distances.[1] There was need for much greater research on the breeding habits and movement of fish. Above all, there was inadequate knowledge of the balance of

[1] 'One of the main difficulties in studying marine pollution and in finding the most acceptable ways of disposing of waste material in the sea is lack of precise knowledge and understanding of how the water itself moves, and how from one place or depth it will run into and with the surrounding water', Dr G. R. E. Deacon, quoted in Dr E. Sibthorpe, 'Ocean Pollution', unpublished paper, p. 15.

nature within the marine environment, and the effect that damage to one species could have on others, and to the complex ecological system existing within the oceans. Here was the first task to be tackled.

A second difficulty was one that was common to other environmental problems: how to promote action by governments and individuals. Even if all the necessary knowledge was at hand, how could self-interested parties, industrial companies, farmers using pesticides, local authorities building sewage works, and national governments concerned with the most rapid possible economic growth, be induced to take the steps necessary to prevent damage to the marine environment, which was often of no direct concern to them? Economic incentives were almost entirely the other way: on the contrary, a company that failed to apply controls of the type required or to find alternative ways of dumping industrial wastes, might be saved enormous expenditure (for example, in the treatment of effluent), which put it at a grave disadvantage in relation to competitors. Even the cliché that those who created pollution should pay for its removal, though it has a superficial attraction, in many cases is neither practicable nor just.[1] Many governments themselves felt that the question was of no immediate concern to them. Those of developing countries, in particular, tended to feel that pollution was caused almost entirely by the rich countries, and that the problem was for those countries to solve: because it was they who were creating the dirt, they should clear it up after them. They certainly did not feel inclined to undertake the substantial expenditure required from their own already inadequate resources to meet the problem. The protection of the environment, in the sea as elsewhere, was basically a social problem and required social action—in this case, action by the international community as a whole. But effective joint action of the kind needed was not easy to induce.

A third problem arose from the interdependence of the different forms of pollution and the complexity of the processes involved. Despite popular enthusiasm, there was no general 'problem of pollution', even in the sea alone. There were a hundred separate problems, arising from a number of totally different and quite unrelated causes: the transport of oil, the need for disposal of sewage outside fresh-water areas, the perpetual increase in industrial

[1] Cf. O. Schachter and D. Serwar, *op. cit.*, p. 1: 'The simple maxim that those who pollute should clean up or pay compensation has only limited utility'.

waste, new types of pesticides, and so on. They entered the marine environment from different sources: from the atmosphere or from rivers, from ships or from sewage works, from mining spoil on the sea-bed or oil-wells beneath it. The polluting agents were of quite different character and chemical composition. They affected different organisms and in quite different ways. Some remained in the sea unchanged for decades or centuries, others were dissolved or degraded relatively quickly. Those responsible were individuals, industrial firms or governments. Each kind of pollution might thus require different measures, applied in different ways, by different people, and organized by different authorities. Yet, at the same time, there needed to be the closest co-ordination among the various measures applied.

A fourth difficulty arose from the delicate balance of the marine ecology. This had the result that the long-term result of any measure that might be proposed was difficult to predict. A measure that might have admirable effects in relation to one pollutant might make matters worse in other ways. Some pollution occurred entirely through indirect effects of this kind. For example, nitrogen (from fertilizers) and phosphates (from detergents), far from harming individual species, usually helped them to flourish as never before. But this in itself, as we saw, could destroy the balance of nature and create new difficulties for other species. Some substances suck up all the oxygen in the water, and so make life impossible for other living organisms. This interrelationship of all species had the effect that the measures taken by one organization to deal with one problem or species could have fatal results for other species elsewhere with which that organization was not concerned.

Yet another problem was political. It arose from the wide disparities in economic and technical capacities among states. Even if the rich countries had acquired the necessary knowledge and techniques to cope—as could happen within two or three decades—this was far less likely to be true of the less developed. And the efforts of richer countries would have little value so long as new pollution was being caused by others, who perhaps lacked either the skill, the resources or the will to tackle the problem. Again, on the sea-bed, as on land, the question arose: Was there a conflict between pollution control and economic growth? And if so, should advance be slowed down, and sea-bed exploitation held back? This was another question which could arouse profound political differences. Finally, because pollutants spread so far and so fast, the problem was a world problem, which could be tackled only at world level, and world decisions were hard to secure.

If international arrangements were required, what type of organization was needed? This was the most difficult question of all. Should there be one single organization responsible for every kind of pollution within the seas, including that from sea-bed exploitation? Or should there be a new sea-bed authority primarily to control sea-bed exploitation concerned only with pollution resulting from this? Should those agencies which had already undertaken responsibility in certain fields continue to perform their own specialized tasks on the basis of their own specialized knowledge? Or should a new UN body be created with responsibilities in this field, with its own assembly and council, like the other specialized agencies (this is what the Secretary-General of the UN seemed to favour later when he called, in 1970, for the establishment of a 'global authority to deal with the problem of the environment'). There were others, especially in the United States and some other developed countries, who were opposed to any such idea, believing that this would make the authority too political, instead of the down-to-earth practical body that was needed: these called instead for the establishment of an authority comparable to OECD, quite outside the UN and representing mainly developed countries, to deal with pollution problems.[1] Finally, how would the money be found: by compulsory levies, as in the agencies; by voluntary pledges as in UNDP; or *ad hoc*? Or should this work be regarded as a form of development, and therefore subtracted from normal economic aid?

So here, as in the economic, legal and military fields, there was disagreement on many questions, when, in November, 1967, the sea-bed issue first began to be discussed within the UN.

[1] This was recommended, for example, in an article by George Kennan in *Foreign Affairs* in 1970.

The History

CHAPTER 5

The Maltese Move

By the late sixties, therefore, a series of problems concerning the sea-bed—economic, legal, military and environmental—had begun to present themselves. Within each individual field a growing consciousness of the difficulties was beginning to arise. But there had been no comprehensive attempt, within the UN or any other international body, to confront the problem as a whole.

A number of private organizations, especially in the United States, had called for such an approach. The semi-official Committee on Conservation and Development of National Resources in the United States recommended to the White House Conference on International Co-operation called by President Johnson in November, 1965, that '. . . if rights are to be granted for resources that are the common property of the world community, then decisions on the allocations of these rights or on the methods of acquisition must be made within the framework of international law. A Specialized Agency of the United Nations would be the most appropriate body for administering the distribution of exclusive mining rights'. The Commission to Study the Organization of the Peace, a liberal and internationalist U.S. organization, in 1966 proposed the internationalization of the sea-bed. The World Peace through Law Organization, an association of international lawyers, in the summer of 1967 proposed the 'issuance of a proclamation [by the General Assembly] declaring that the non-fishery resources of the high seas, outside the territorial waters of any state, and the bed of the sea beyond the continental shelf, appertain to the UN and are subject to its jurisdiction and control'. A book financed by the Brookings Foundation, *Financing the UN System*, by K. G. Stoessinger and others, considered, among other things, the possibility of deriving UN revenue from the licensing of exploitation within the sea-bed. Finally, at the Pugwash conference of international scientists in 1967, a U.S. scientist, Alexander Rich, and a Soviet scientist, Academician V. A. Englehardt, joined in proposing that the mineral resources of the sea-bed should be administered by the UN to provide funds for economic development.

Others, even if not necessarily particularly favourable to internationalization, or UN control of the sea-bed, became increasingly conscious of the dangers of unrestricted competition leading to a new Klondyke within the sea-bed. Even the two super-powers, who were likely to be most resistant to internationalization (as they could expect to gain most from independent national action), felt obliged to take some account of this. In July, 1966, President Johnson declared: 'Under no circumstances, we believe, must we ever allow the prospects of a rich harvest of mineral wealth to create a new form of colonial competition among the maritime nations. We must be careful to avoid a race to grab and hold the lands under the high seas. We must ensure that the deep seas and the ocean bottom are, and remain, the legacy of all human beings.' This last was a significant phrase which received a pronounced echo in the subsequent debates within the UN, of a kind that its author may not have expected. Even the Soviet Union, which has been consistently hostile to any move to give increased powers to the UN, in this or any other field, let alone to give it an independent source of revenue, began to be aware that pressures for internationalization were beginning to build up. Soviet reaction was to call for a strengthening of the role of the IOC, in its eyes the least dangerous and supranational of international institutions, because the most technical. In February, 1967, the Soviet representative of the Consultative Council of the IOC proposed that it should establish a working group to draft conventions on the exploration and exploitation of the mineral resources of the sea (such conventions would not of course have been absolutely binding but would have depended on the agreement of each state).

The UN too had begun to take some account of the prospect that exploitation in the sea-bed might shortly begin. On March 7, 1966, the Economic and Social Council (ECOSOC) passed a resolution which requested the Secretary-General to 'make a survey of the non-agricultural resources of the sea beyond the continental shelf and of the techniques for exploiting' them and to 'identify those resources now considered to be capable of economic exploitation, especially for the benefit of the developing countries'. Already, therefore, the idea was established that sea-bed resources should be exploited for the benefit of developing countries. Finally, when the whole issue of the sea-bed was first raised at the 1967 Assembly, the Secretary-General made the personal proposal that he should examine various alternatives, 'including the advisability and feasibility of entrusting the deep-sea resources to an international body'.[1]

By 1967, therefore, there was a rising concern in the UN and elsewhere about some of the possible implications of sea-bed development. Various specialized bodies had become active. Resolutions had been passed on specific aspects. But there was very little general awareness of the problem among the general public, or even within the UN, and there had been no attempt at all to mobilize any international body to confront the problem as a whole.

It was left to Malta, one of the smallest members (but one which was intimately concerned with the problems of the sea) to arouse the UN to action on this subject. On August 18, 1967, a month before the beginning of the 1967 General Assembly session, the Maltese delegation formally proposed the inclusion of a supplementary item on the Agenda, to be entitled 'Declaration and Treaty concerning the reservation exclusively for peaceful purposes of the sea-bed and the ocean floor underlying the seas beyond the limits of present national jurisdiction, and the use of their resources in the interests of mankind'. This was accompanied by a memorandum, setting out the reasons why Malta felt the question should be considered, and calling for 'immediate steps' to draft a treaty on the subject.

When the General Committee met, it agreed to accept the item. However, this immediately aroused consternation amongst Latin-American delegations. These were concerned that in the form originally suggested the item could lead, perhaps fairly soon, to the formulation of a declaration or treaty on the sea-bed, which might lay down the maximum permitted limits of territorial waters or continental shelf, or both. Because, as we have seen, some of them had declared limits well beyond those regarded as

[1] Note by the Secretary-General, A/C1/952 of October 31, 1967.

reasonable by most other states, this could have placed them in a position of great difficulty. Powerful representations were therefore made to the Maltese Ambassador to the UN, Dr Arvid Pardo, both privately and in the Assembly itself, imploring him to withdraw the item.

It was important to Malta to retain the goodwill of the Latin-American states, who control more than 20 votes within the Assembly. So Malta finally agreed to amend the item slightly to 'Examination of the question of the reservation exclusively for peaceful purposes . . .', so not prejudicing in advance the question whether any declaration or treaty should be the final result. The item was allotted by the General Committee to the First Committee, which deals mainly with disarmament and security matters.

On November 1, 1967, Dr Arvid Pardo was invited to introduce the item. He then embarked on a long, learned and eloquent disquisition, lasting over three hours—a long time, even for UN speeches—on all the main aspects of the question. It can safely be said that the members of the First Committee, accustomed to listening over and over again to well-worn arguments on well-worm themes repeated interminably from year to year, have rarely listened with such fascination as they did in hearing the wonders of the deep explored by the learned doctor with such erudition and force.

He first explained the nature of the different types of problem that were beginning to emerge in various fields in relation to the sea-bed. He recounted those types of economic exploitation that were already being carried out, especially through off-shore oil drilling, and the possibilities of the exploitation of manganese nodules. He recalled the advances in undersea technology and in man-in-the-sea techniques, which could bring the exploitation of these undersea resources so much nearer. He painted the dangers of the possible military uses of the region, and of competition for national appropriation. He described the pollution problem. And he finally proposed that the UN should take action on the sea-bed similar to that which it had already taken in relation to outer space: thus the Assembly might pass a declaration that the sea-bed and ocean floor were 'the common heritage of mankind' (an ingenious echo of President Johnson's phrase), on the lines of the Declaration on the Uses of Outer Space which had been passed in the 1963 General Assembly (subsequently elaborated in a Treaty). This would set out the general principles to be followed in the use and exploitation of the sea-bed, and would call on member states to refrain from extending their national claims until a common decision had been reached about the limits of the rights of coastal states. At the same time, 'a widely representative but not too numerous' body should be established to consider the political, the economic and other implications of setting up an international régime for the sea-bed, and to draw up a comprehensive treaty designed to 'safeguard the international character' of the sea-bed and define the limits of national rights. An international agency could then be set up to ensure that national activities on the sea floor conformed to the principles contained in the treaty and perhaps administer the area generally.

There is no doubt that the Maltese initiative, and Dr Pardo's speech in particular, made a profound impact on the Assembly. In the delegates' lounge, the spacious bar and smoking room where the delegates congregate between meetings, conversation tended to centre on the Maltese initiative. In the innumerable and interminable cocktail parties, representatives would ask one another how their governments would react to Dr Pardo's proposals. There was a general feeling that the UN had here become involved in a new subject, of profound importance but great complexity and fascination, which would command the attention of delegates and officials for many years to come.

But there were very many conflicting opinions, both about the merits of the proposal (in so far as the issue was understood), and about the best way of handling it. There were some from the rich countries who were profoundly suspicious of any interference by the UN in the exploitation of the newly discovered regions beneath the sea. Dr Pardo himself quoted the words of the U.S. congressman who demanded: 'First, . . . why did the Maltese Ambassador, Arvid Pardo, make this premature proposal? Second, who put the Maltese government up to the proposal? Are they perhaps the sounding board of the British? Third, and most of all, why the rush? It is my conviction that there is no rush. . . . There is little reason to set up additional unknown and additional legal barriers which will impair and deter investment and retard exploration in the depths of the sea even before capabilities and resources are developed.'[1] This theme, that impetuous and excessive UN interference in the question would

[1] *Congressional Record*, September 28, 1967.

deter private economic exploitation, was one that was to recur very often in the speeches of some countries in the years to come.

Apart from fundamentalist objections of this kind, there were others who for different reasons were suspicious of action as rapid and decisive as the Maltese delegate had called for. The fact that the entire subject was so new, at least in the UN context, provided some justification for demands that much more time was needed for reflection before any declaration or other statement of principles was passed. Even the delegates of many poor countries were unwilling to be rushed into firm commitments before they had had time to consider exactly where the long-term interests of their own countries lay. And the representatives of the Latin-American countries, in particular, were profoundly hostile to any attempt to rush the question in a way which might reopen the sensitive issue of the limits of territorial waters and continental shelf on which they knew their own position to be unorthodox, to say the least.

Formal consultations on what should be done began immediately. A wide variety of views inevitably emerged. To some extent these reflected the conditioned reflexes of particular groups on subjects of this kind. The Communist states had a long-standing aversion to the establishment of new UN committees or bodies of any kind, above all of any which might eventually acquire substantial authority in particular fields. They accordingly called for further detailed studies before making any decision, even about the way the question itself should be handled in the future, let alone agreeing on substantive principles. The Secretary-General should simply be asked, in co-operation with the IOC and other bodies, to prepare a report describing the present state of knowledge on possible exploitation of the area and recording the activities of UN bodies.

The representatives of western developed countries were also cautious, though anxious to give the impression of being 'positive' (a somewhat acrobatic posture frequently adopted by such powers within the UN). They pointed out that this was the first time the question had ever been considered within the UN. The subject was vast and complex. It would be a mistake to be precipitate. All that was necessary was to ask a committee of experts, or an *ad hoc* intergovernmental committee, to look into the question and to report at the next Assembly.

The Latin-Americans, too, were guarded in their response. As developing countries, with little hope of being able to undertake exploitation of the deep sea-bed through their own resources, they might have been expected to give strong support to an initiative designed to preserve the benefits of the area for the international community as a whole. But their thinking, and especially that of the Pacific countries and Brazil, which in turn influenced the others, was largely dominated by concern that nothing should be done that might prejudice their claims to extensive territorial waters or continental shelves. These too, therefore, warned of the danger of hasty action.

Against these three groups was a fourth and larger group, consisting of Malta and a number of other developing countries, together with a few well-wishers, such as Sweden. This group had in common their anxiety over the urgency of the question. They were concerned that, unless some machinery was established soon, the future of the sea-bed might be permanently prejudiced by the action of governments making claims or companies beginning to exploit the area beyond the limits of national jurisdiction. Sweden demanded therefore that, even if no declaration could be passed at that Assembly or any permanent machinery established, at least a call should be made to all member states to avoid any action or claims which could result in an alteration of the legal *status quo*.

Bitter and protracted discussions took place behind the scenes. All the various viewpoints were put forward with varying degrees of passion and intensity. Ultimately, as is usually the way in the UN, something of a compromise was reached. The Assembly was to set up an *ad hoc* committee to study the 'scope and various aspects' of the item. The committee would study the whole question raised by Malta and report to the next Assembly with an account of the scientific, technical, economic, legal and other aspects, as well as of past and present activities on these questions by UN bodies. It would also provide for the Assembly 'an indication regarding practical means to promote international co-operation in the exploration, conservation and use of the sea-bed', having regard to the views put forward at that Assembly. The new Committee was ultimately established with 35 members, roughly 20 developing and the rest developed, including 5 Communist states. It did not, however, represent any land-locked developing countries, a decision which caused justifiable complaints from Paraguay and Nepal. This was a question on which there was an obvious and radical divergence of interest between coastal and non-coastal states, and it was a glaring departure from the UN's normal practice on such questions that no representative of the non-coastal states was included in the Committee. This was the first indication of the battle between these two groups that was to come.

The *Ad Hoc* Committee

When the new Committee met, in March, 1968, it decided to establish two Working Groups: on technical and economic questions, and on legal questions. The Working Groups were to consider the main problems in their own field and submit reports to the main committee, which would review them at a later session. The permanent delegate of Ceylon, Mr Amerasinghe, was elected chairman of the main committee, a position he has occupied with distinction ever since.

Most of the delegations at this time, and many of their governments, were wholly ignorant of the subject: they did not even know, sometimes, what continental shelves they themselves possessed, still less what resources they might contain. The first step was, therefore, to call for massive documentation from the Secretariat on a number of technical, legal and other questions. Meanwhile, in most countries the protracted and complex process of interdepartmental consultation and consideration, which was necessary within every member government on this complex question, had in most cases barely begun. The whole question was one of huge complexity, in which a number of different interests, private and public, were at stake. Different government departments, concerned with different objectives, were involved, sometimes in conflict with one another. The department, and the private interests, concerned with fuel supply might have quite different views of the desirable type of régime from those concerned with, say, fishing, scientific research or defence. It is thus scarcely to be wondered at that most delegates at this time were not ready to put forward any substantive views on most of the questions which arose during the early meetings.

When the Working Groups met, certain clear-cut differences of view already began to emerge. Within the legal working group, there were those who felt that the law of the sea-bed, including the legal definition of where it began and where the continental shelf ended, was still to be made through some international act of decision; and those who felt that existing instruments of international law already covered such matters (and, some held, protected a general freedom of action), so there was no need for any new definition. The former was the view of Malta, Sweden and a number of other countries, many of whom favoured calling a special conference on the law of the sea, or even an attempt by the Committee or the Assembly to resolve the question through a resolution. Against these were most of the Latin-Americans, who held strongly to the view that maritime limits had already been decided (by unilateral action) and that any attempt to review that issue now would be an infringement of national sovereignty and could lead to endless international conflict. The rich countries were equally divided. Some took the view that it would be counter-productive to attempt to call a new conference on the law of the sea, since there was little reason to think that it could reach any more agreement than that of 1958, but that decisions on a new international system might be required. Others wanted a conference but only to define the territorial sea and other undecided questions. And some, such as Britain, urged that one could not in any case reach a decision on the limits of the shelf until one knew what kind of régime would be set up beyond the shelf (though it was equally logical to urge that one could not decide the régime until one knew the area it was to cover).

There was another conflict of views over the type of régime which should eventually be aimed at. The developing countries as a whole were naturally in favour of the most international system possible. Just as the poor within states may hope to gain from a system of nationalization, which may share the wealth of the nation among all its people, so the poor of the international community naturally supported the internationalization of resources that had never belonged to any particular nation so that the benefits of these could be shared among the people of the world as a whole, especially those least well-off. They united around the concept, first stressed by Malta in her ingenious paraphrase of President Johnson, of the sea-bed as the 'common heritage of mankind'. For this served to emphasize that these were resources which belonged to the world as a whole, could be exploited only with the permission of, and under principles determined by, the international community, and should be used for the benefit of all countries, above all developing countries.

The rich countries on the other hand tended to utter dark warnings about the dangers of deterring investment in the development of undersea resources. Attempts to establish a complex system of international control, or to demand heavy royalties for international purposes, it was suggested, might only succeed in ensuring that no development took place at all. A few others, such as Malta and Sweden, warned of the dangers that would result from an uncontrolled national scramble for the resources from the area. So there were still many different views on the course that should be pursued.

The Principles

All these differences came to a head over the attempt to formulate a set of principles to govern the exploitation of the sea-bed in the future. This, as we have seen, had been one of the main demands of Malta the previous year. When the *Ad Hoc* Committee reassembled in Rio in August, 1968, at the invitation of the Brazilian Government, it was now mainly to this task that it devoted itself.

The disagreement was mainly between 'minimalists' and 'maximalists': between those who wanted only a set of principles that everyone could accept with little difficulty—which meant that they would utter very little other than pious platitudes; and those who wanted a set of much more ambitious and controversial proposals with specific content—which meant that they might not be accepted by all and might have to be forced through by a majority vote.

Differences on this point reflected underlying differences of interest, above all between the rich countries and the poor. The developed countries wanted a set of harmless statements of goodwill, expressed in the most general terms, committing them to very little other than the traditional obligations to preserve existing uses of the seas, and leaving the future system of exploitation and legal rights largely open and undecided. The developing countries, on the other hand, wanted a set of principles which would carry the situation beyond where it already stood; which would establish, if possible, the principle of international ownership of the sea-bed resources, and at least the need for some system of international control over their exploitation.

There were, of course, also, many differences on specific issues.

The substance of these is considered in Chapter 8 below. Two rival sets were put forward by the United States, on the one hand, representing the minimalist position, and by a number of developing countries, putting the maximalist view. Eventually a middle road was put forward in another set of principles, originally proposed by the United Kingdom, but supported later by other developed countries. This was an attempt to find the lowest common multiple of the other two sets of principles put forward. It went somewhat beyond the extreme nationalist position of the Soviet Union and other Communist countries, and the extreme private-enterprise position of the United States. But it went nowhere so far as the extreme internationalist position of the developing countries. The seven proposed principles were as follows: 'there is an area of the sea-bed which lies beyond the limits of national jurisdiction; a precise boundary should be agreed for this area; an international régime governing the exploitation of resources of this area should be agreed upon; no state may claim or exercise sovereign rights over any part of this area and no part is subject to national appropriation; exploitation of the area shall be carried on for the benefit of all mankind, especially the developing countries; the area shall be reserved exclusively for peaceful purposes; and activities in this area shall be conducted, in accordance with international law, including the UN Charter'.

Most of these were relatively uncontroversial. But this was only because they were also relatively vague. The statement went beyond the U.S. and Soviet position in speaking of the need for a régime; but this was a word that could be interpreted in innumerable ways. It stated that a precise boundary should be agreed; but it did not say when or how—still less what it should be, the really essential point. It said that no state could claim sovereign rights over the sea-bed; but did not say whether this meant free exploitation by anybody or controlled exploitation under an international authority. It said that exploitation should be for the benefit of all mankind; but did not say whether this meant a need of maximum production as quickly as possible, or that exploitation should be under strict control and against the payment of large international royalties. It said that the area should be reserved only for peaceful purposes; but did not say what activities this prohibited. It said that activities should be in accordance with international law; but did not attempt to indicate the rules this presupposed. For most of the developing countries, and therefore for the majority of those represented, it thus did not go nearly far enough in defining the rules to govern exploitation in the future. For a few of the developed, including the Communist countries, the statement went too far.

Because of such differences it was not possible to agree on a set of principles that year. The *Ad Hoc* Committee had to be content with drafting a report for the Assembly in relatively simple terms. It was able to report general agreement on a proposal put by the United States that there should be an International Decade for Ocean Exploration; and on another proposed by Iceland, that a report by the Secretariat should be undertaken on the means of preventing pollution in the exploitation of the sea-bed. More important, it was able to record general support for a Belgian proposal that, at the forthcoming Assembly, the *Ad Hoc* Committee should be replaced by a permanent Committee to continue discussion of the sea-bed problem.

The 1968 Assembly

At the 1968 Assembly, many of the same points of disagreement which had arisen in the Committee came to the surface once more among those not represented on the Committee. A resolution to establish a permanent Sea-bed Committee was put forward, generally neutral in its wording, to attract the widest possible support. It thus had to stress such concepts as 'international co-operation', a phrase beloved by the Russians, to which nobody would reasonably object but which could mean all things to all men. The resolution was supported by a broad group of countries, and represented thus in a sense the lowest common multiple of the Assembly. But it could not satisfy some developing countries, which were looking for a more radical solution. These proposed an amendment under which the Committee would specifically have examined 'the question of establishing in due time an appropriate international machinery for the exploration and exploitation of this area . . . and the use of these resources in the interest of mankind, especially those of developing countries. . . .' This phrase, particularly the reference to 'international machinery', was quite unacceptable to many of the developed countries, above all to the Soviet Union and the Communist states. Such countries felt that the inclusion of this phrase would have totally prejudiced the outcome of the proposed Committee's discussions in favour of such 'machinery', which they abhorred. Eventually, the sponsors were prevailed on to make two important changes. On the one hand the proposal on 'machinery' was included in a separate resolution from the main one establishing the Committee (so that separate votes of dissent could be recorded); and on the other, it called on the Secretary-General to prepare a report of the 'advisability' of establishing such machinery, which was a little more non-committal.

There was still some hope it might be possible to reach agreement on a statement of principles during the Assembly. Various versions, new and old, were proposed. But all of these raised the same points of disagreement which had led to deadlock at Rio. Eventually, the Maltese representative proposed, and the other sponsors agreed, simply to submit all the resolutions that had been proposed concerning principles to the new committee that was now to be established. Thus the disagreements, as they could not be dissolved, were merely removed to another forum.

The final point on which dispute occurred was the composition of the new permanent Committee. Such questions are always delicate problems within the UN. In this case particularly, there was a realization that the Committee to be established would have a unique importance, being concerned as it was with the future control of a large part of the world's resources. It was likely to come up with recommendations that could have a profound importance for the interests of every member state and the distribution of the world's wealth in the future. Since interests differed from state to state according to geographical situation, every member wished to have its own voice in the proceedings. A process of intensive bargaining therefore occurred. It was finally agreed that the new Committee should have 42 members, with a system of rotation to enable maximum participation. Eventually, on this basis, after much wrangling and recrimination, the composition of the Committee was finally agreed when the Assembly was only just about to break up. The size of the Committee was already probably too big for efficient activity; but it was nothing like so big as it was eventually to become.

The resolution finally passed proposed that the new Committee should 'study the elaboration of the legal principles and norms' that would promote international co-operation in the exploration and use 'of the sea-bed and would ensure exploitation for the benefit of mankind': in other words, the Committee should continue with the attempt to arrive at agreed principles, and to clarify international law on the sea-bed. It should also study 'the ways and means of promoting the exploration and use of the resources of the area, and of international co-operation to that end'; in other words, it would begin to consider the substance of the régime to be established. An integrated long-term programme of ocean exploration, including the international decade of ocean exploration which the United States had demanded, was to be launched. The Secretary-General was called on to prepare a report on the means and principles for the conservation of the living and other resources of the sea-bed, as Iceland had proposed. Finally, the Secretary-General was to prepare a study on 'appropriate international machinery for the promotion and exploration and exploitation of the resources of the area and the use of these resources in the interests of mankind'. The Communist countries voted against this proposal: the only substantial point of disagreement among all the resolutions passed.

So, for the first time, a permanent UN body was established to consider the problems of the sea-bed. Some parts of the question began to be considered elsewhere. Some of the disarmament questions had already been raised by this time in Geneva and were to be considered increasingly intensively there from this point on. The pollution problems were the concern of a number of bodies within the UN system already (IMCO, the IOC, FAO and others). From this time they were also being increasingly considered in the context of the 1972 UN Conference on the Environment (which the 1968 Assembly also approved in response to a proposal from Sweden). The legal questions too, it was hoped by some, might soon be looked at by an international conference on the law of the sea, perhaps called especially for the purpose, rather than by the new Committee.

The new Committee was thus left mainly with the central core of the problem in its hands: to whom did the resources of the deep sea-bed belong, and how were they to be exploited?

CHAPTER 6

The Disarmament Debate

The military strategists are perhaps always quicker on the draw than the disarmers. Certainly they were quicker in becoming conscious of the strategic potentialities of the sea-bed. As we saw in Chapter 3, there had been a growing interest, through most of the sixties, among strategic writers and military staffs in the enticing military possibilities of that region. There was for long no corresponding concern about the military dangers among the public. It was not until after the Maltese initiative at the UN that the need for measures to control those dangers began to be widely recognized.

When Ambassador Pardo raised the sea-bed issue in the autumn of 1967, he had laid special stress on the disarmament aspects. His original intention, as we saw, was that a Declaration reserving the area 'exclusively for peaceful purposes' should be passed at the same Assembly. He described, in vivid terms, the possible strategic uses of the underwater areas. He warned that there were 'grave considerations of security and defence interests that impel the major powers to appropriate areas of the ocean for their own exclusive use'. He pointed out that the importance of the sea would increase rather than decrease in the age of the nuclear submarine. Moreover, there were further and more dangerous possibilities: 'What could be more attractive in the era of multiple-warhead ballistic missiles, capable of overwhelming defences and destroying land-based missile sites, than to transfer offensive and defensive capability to the sea, an environment highly resistant to the pressures of nuclear attack.' Fixed military installations on the ocean floor might be found useful for many purposes. There could be a race to occupy the most accessible strategic areas. He therefore called for 'credible assurances that the sea-bed and ocean floor will be used exclusively for peaceful purposes'.

The concern to prevent a new arms race in an area hitherto immune was widely shared. But the attitude of most powers was, as always, largely determined by their own strategic interests. These differed radically.

The Soviet Union, as a power which had concentrated a great part of its naval strength in its large and powerful submarine fleet, had a strong interest in ensuring that any agreement for demilitarizing the sea-bed should prohibit the use of anti-submarine detectors and similar devices which might reduce that power; but that it should not affect the freedom of submarines and other vessels in the waters of the seas themselves. The United States, on the other hand, whose naval power was already threatened by the Soviet submarine fleet, had a diametrically opposite interest: that an agreement should not interfere with her capacity to track and detect submarines through such devices, which were essential to her in countering the Soviet naval threat. Britain and other European countries, also highly vulnerable to the Soviet submarine fleet, had an interest generally similar to that of the United States, but had no developed sea-bed capacity themselves. The smaller nations, having no capability at all in underwater technology but apprehensive over the super-powers' increasing activity, mostly wanted to see a total and complete prohibition of every form of military activity in the sea-bed, especially close to their own coasts.

Attitudes were complicated by other factors not exclusively military. Those countries, especially Latin-American, which claimed wide territorial waters and shelves were insistent that any agreement on demilitarization should not prejudice their claims. For example, they feared that demilitarization in waters close to their shores might involve inspection taking place in such areas, which they claimed they alone should control. Each country had to weigh the advantage of preventing other countries from taking *offensive* measures in the sea-bed against that of retaining freedom to undertake its own military activities in the same areas, even if only for defensive purposes. Some countries, such as Canada, appeared to attach greater importance to reserving their own freedom in the defensive field than to prohibiting the activities of their enemies in the same areas.

These varying attitudes were quickly reflected in the debates. In March, 1968, the Soviet Union proposed the total demilitarization of the sea-bed, with a prohibition of all military uses of any kind. It put down a resolution within the *Ad Hoc* Committee calling on the Eighteen-Nations Disarmament Committee (ENDC) to 'consider as an urgent matter the question of prohibiting the use for military purposes of the sea-bed and the ocean floor beyond the limits of territorial waters of the coastal states'. The United States, on the other hand, put down a formal resolution requesting the ENDC to 'take up the question of arms limitation on the sea-bed and ocean floor, with a view to defining those factors vital to a workable, verifiable, effective international agreement which would prevent the use of this new environment for the emplacement of weapons of mass destruction'. And in the ENDC the U.S. delegate proposed that the conference should take up the question of 'arms limitation' in the sea-bed. The difference in the terminology employed was by no means accidental.

So the battle lines were set, which were to remain unchanged for the next year or two. One super-power was calling for prohibition of all military uses (whether or not this was verifiable); the other for the prohibition only of weapons of mass destruction.

In the ENDC, in March, 1969, the Soviet Union formally submitted the text of a draft Treaty for the total disarmament of the sea-bed.[1] The treaty prohibited the placing not only of nuclear weapons of mass destruction, but also of all 'military bases, structures, installations, fortifications, and other objects of a military nature'. Any installation that was established and might be challenged by another nation as having a military potential 'would be open on the basis of reciprocity to representatives of other states'. The proposed treaty would cover the whole of the ocean floor 'beyond the 12-mile maritime zone'. By omitting any reference to 'territorial waters' in this way the Treaty avoided the problem that arose because of the widely differing extent of national claims to territorial waters.

The U.S. delegate, Mr Gerard Smith, strongly argued against total demilitarization. He pointed out that 'the existence of submarine fleets requires states to take action in self-defence, such as establishing warning systems within the sea-bed. At the same time much useful scientific research was supported or carried

[1] UN Doc. A/7741.

out by military personnel using military non-weapon equipment'. Inspection of the entire sea-bed area could not practicably be undertaken. Therefore 'complete demilitarization of the sea-bed, would in our judgement, be simply unworkable and probably harmful'. The most sensible course was to concentrate on a ban on the emplacement of weapons of mass destruction, whether nuclear, chemical, biological or radiological. This was where the real danger lay, he claimed. So the United States, too, introduced its own draft treaty on these lines.

Only after the two main super-powers had had their say did other delegates begin to join in the disarmament debate on any scale.

The Three Main Issues

There were really three basic issues which were in dispute here. The first, and the most important, already highlighted in the different views of the United States and the Soviet Union, concerned the *scope* of any treaty reached: should all or only some military activities be prohibited? The second concerned *the extent* of a treaty; how large an area of the sea-bed was to be covered by its provisions? And the third concerned its *enforcement*; how was verification that its provisions were being applied to be carried out?

On its *scope*, the bulk of opinion was with the Soviet Union. Representatives of the non-aligned states all made it clear that they supported a comprehensive treaty banning all military activities of any kind. Sweden, Ethiopia, the United Arab Republic, Brazil and others all spoke in this sense. But it is not certain that all delegates had understood the significance of underwater detector devices, or had considered whether a ban which prohibited these would really make the marine environment as a whole a safer or less safe place. The feeling was simply that there was a danger that the arms race would spread to a new area: any suggestion that demilitarization should be less than complete must thus conceal a sinister attempt to extend military activities into the sea-bed.

Only a few western countries, such as Britain and Italy, took a different line: while further measures should be considered 'on their merits', the first step should be to build on the degree of agreement which already existed by banning nuclear weapons and weapons of mass destruction, as the United States proposed.

The United States argued that, if all objects of a military nature were to be banned, verification problems would be insuperable. Since small objects in the sea-bed could not be detected in the first place, the parties would have to rely on good faith alone, which might not always be present. Against this, the Soviet delegate argued that no real distinction should be made between offensive and defensive weapons. If some types of equipment were excluded, military activity would be legitimized, and the sea-bed would become a new arena for the arms race. The use of military personnel for military research and the 'emplacement of communications devices or beacons' need not be banned. But, the Soviet delegate argued, sea-bed weapons 'designed to strike at ships and to disrupt sea communications' were as much in need of control as those designed to strike at the territory of other states.[1]

On the question of the *area* that the treaty would cover, disagreement was less violent. The U.S. draft related to the whole sea-bed outside a three-mile zone. The Soviet draft treaty covered the whole area except the 12-mile maritime zone. Each of these corresponded to the limit of territorial waters which each country acknowledged. The U.S. delegate argued that the advantage of the U.S. proposal was that it would cover the largest possible area, and that it would avoid having an area which for some countries would lie between the edge of their territorial waters and the area covered by the treaty. In general, there was most support for the idea of a twelve-mile limit. Many countries recognizing a twelve-mile limit were reluctant to see control measures take place within their own waters. Japan proposed that there should be no excluded area at all. But she was supported by virtually no one: such a system would create enormous difficulties for a coastal state, not only by preventing it taking its own defensive measures, but by allowing inspection close in to its own shores.

There were two other difficulties, however, connected with the area to be covered. Brazil, representing many Latin-American countries, was concerned that inspection should not take place anywhere in the continental shelf without the consent or even the participation of the coastal states. This desire for a special right of control for the coastal state was a long-standing Brazilian

[1] ENDC, PV 411.

demand which we shall encounter later in the economic sphere. But the proposal of course immediately raised the key point concerning the limits of the continental shelf and the extent of the coastal state's powers there. The majority of states were unlikely to accord such sweeping rights of control in the shelf as Brazil demanded, above all if states could claim virtually any limits they pleased, as some Latin-American states had done.

There was another point, raised by Canada and Italy. Should a country be able to take *defensive* measures beyond the proposed limits? As countries having territories divided from the mainland by large stretches of water, both were concerned that the right to take defensive measures on the sea floor should extend well into the ocean. Canada proposed that, whatever the demilitarized area, there should be a 200-mile 'defensive zone', in which, even if a prohibition existed for everybody else, the coastal state should be allowed to undertake limited defensive measures, as might be provided in the treaty. In particular, the coastal state would have sole responsibility for policing the agreement in this area. Similarly, Italy questioned whether the concept of an area defined only by distance from the coast, as laid down in the Soviet and U.S. drafts, 'would be sufficient to guarantee the security of coastal states and more particularly that of countries whose geographical configuration was characterized by a very large sea frontier'. She proposed therefore, instead, the 200-metre isobath or the twelve-mile line, whichever was greater: in other words, much of the shelf would be included in the defensive zone. Sweden proposed a kind of three-tier system as a compromise: weapons of mass destruction would be prohibited from three miles from the coast as the U.S. draft proposed; a more comprehensive ban would cover the area outside the twelve-mile zone as the Russians wanted; and there would also be a defensive zone of 200 miles in which the coastal state had special rights.[1]

The final controversy concerned the method of *verification*. This presented a somewhat ironic spectacle. Almost universally, in discussing disarmament measures the United States has demanded inspection as essential to verification, and has refused to agree to some (for example, a comprehensive nuclear test ban) because of the lack of such a provision. The Soviet Union has, equally invariably, taken exactly the opposite position, refusing to

[1] Nigeria also wanted a defensive zone, but of 50 miles only.

allow any kind of inspection on its own territory, and denouncing this as an unnecessary violation of sovereignty.

In the sea-bed discussions however (where its own territory was not at risk) the Soviet Union insisted strenuously on the need for 'free access' to any object which might be suspected as being military, and for inspection of it thereafter if necessary. Its draft treaty provided that all installations would be open to verification 'on the basis of reciprocity'. Any party would have the right to approach the facilities of any other party and to inspect it on demand (this was somewhat similar to the provisions of the Antarctic Treaty). The United States in this case strongly resisted the right of inspection, arguing that to accord these rights was unnecessary. It could cause disruption in the economic exploitation of the sea-bed if demands for inspection were unreasonably made. It could, moreover, be the means of acquiring economic secrets by inspection of oil or other installations or even military secrets (such as those surrounding detector devices). The United States proposed that, in the case of a challenge by one country against another, the only obligation on the second party would be to 'consult and co-operate' on the means of resolving the problem. This was regarded by most other countries as a somewhat inadequate assurance.

There was another problem: who was to be responsible for verification? The Soviet provision for verification 'on the basis of reciprocity', and the U.S. proposal to 'consult and co-operate', had in common that they would mean that in practice only the super-powers themselves would take part in inspection: by inspecting each other. Both countries expressly denied the need to set up any special international body to be responsible for controlling the agreement. Most other countries argued strongly either for the establishment of an international control body (Sweden, Italy, India); or for the right of any power to request that another country with the necessary sea-bed technology should assist them in inspecting suspicious events (Brazil, Canada, Nigeria); or for *ad hoc* agreements among groups of powers, by which all within a region would assist each other in inspection (Mexico). Even two Communist countries, Yugoslavia and Rumania, supported the call for an international control authority. Sweden wanted the coastal power to have not only the right but the obligation to undertake verification in its own maritime zone. Finally, Canada made the ingenious proposal that if an offence was discovered, the offending state would pay the cost of inspection, while if not the complaining state would pay.

The two super-powers resisted these proposals. The Soviet Union was, as always, strongly opposed to the establishment of any new international body, especially one that might acquire such wide powers. The United States pointed out that all states could undertake some forms of verification. No large-scale weapons could be emplaced without some kind of surface activity, and any country with ships or aircraft could observe these, challenge them if necessary and even arrange for divers to descend to inspect at closer quarters. In this sense, the position of less advanced countries would not be much different from that of the more advanced. The U.S. treaty anyway contained a provision for review, which would enable the signatories to consider whether the inspection arrangements needed to be revised, as well as whether new categories of activities needed to be prohibited. This would be far better than seeking to establish a new international agency. As the U.S. delegate put it: 'we do not see how an international organization could now be established, staffed and equipped to perform tasks concerning which we have so little experience.'

All of these were genuine difficulties, which face any agreement for disarmament in the sea-bed. Many of them remain. There were, however, some rapid changes of attitude in the following months. When the first meetings ended in July, 1969, it was known that during the recess that autumn there might be unofficial consultations. There was some faint hope that something might come of these. Few members of the conference, however, can have expected the rapid and astonishing change in positions which then took place.

The U.S.-Soviet Treaty

When the conference resumed its session at the end of September, 1969, the two super-powers produced the text of a *joint* draft treaty, already agreed between them. It was now commonly sponsored by them.

A short examination quickly revealed that, on almost every point, the Soviet Union had decided to accept the U.S. view. On the main issue in dispute, its scope, it accepted that the treaty should apply only to nuclear weapons and weapons of mass destruction. It accepted verification through an agreement merely 'to consult and co-operate'. And on one or two lesser points, the U.S. position had again been adopted.

Under the proposed treaty, the parties were to undertake not to emplace on the sea-bed 'beyond the maximum contiguous zone provided for in the 1958 Geneva Convention on the territorial seas and the contiguous zone [i.e., more than twelve miles from the coast] any objects with nuclear weapons or any other types of weapons of mass destruction as well as structures, launching installations, or any other facilities specifically designed for storing, testing or using such weapons'. Nor would they help or encourage others to do so. To ensure observance, the states party to the treaty would have the right to verify the activities of other parties in the sea-bed (without 'interfering with those activities or infringing on recognized rights'). If they found a suspicious activity, both parties were to consult, and to co-operate 'with a view to removing doubts concerning the fulfilment of the obligations' assumed under the treaty. Signatories were allowed, as a final recourse, the right to withdraw because of 'extraordinary events related to the subject-matter' of the treaty; a final sanction that had been granted in the non-proliferation Treaty.

One slight concession to those who wanted a more comprehensive ban was contained in the preamble to the treaty. This included the clause, 'Convinced that this treaty constitutes a step towards the exclusion of the sea-bed, the ocean floor, and the subsoil thereof from the arms race, and determined to continue negotiations concerning further measures leading to this end. . . .' So there was some kind of promise to discuss wider measures later. Again, under Article 4 it was provided that any state could propose amendments at a subsequent stage to extend its scope. But amendments had to be approved by a majority of signatories, including all nuclear powers which were signatories. This meant that any signatory nuclear power would have a veto on subsequent amendments, including any designed to extend the treaty's scope.

Understandably, the new joint draft caused some bewilderment among the members of the conference. The majority which had supported the Soviet view that the treaty should ban all military uses of the sea-bed, whether nuclear or conventional, were somewhat dumbfounded to find that the main protagonist of this viewpoint had now completely abandoned that position. Sweden quickly sought a more specific commitment, in the main body of the treaty, that negotiations would continue for complete demilitarization: a demand that received widespread support from others. Secondly, there was opposition to the proposed control arrangements. Canada called for the right of a challenging state to demand verification of suspicious events by inspection, if necessary, though without interfering with the activities involved; in addition, each state should undertake to grant 'such access as may be required' to make inspection possible (as the earlier Soviet draft had demanded). Thirdly, many states resisted the veto on amendments accorded to the nuclear powers in the U.S.–Soviet draft. A fourth point criticized by a number of states, especially the Latin-Americans, was that no definite right had been accorded to the coastal state to be consulted over any inspection or verification carried out in or above the continental shelf. The United States and other developed states felt that to grant this could prejudice their general position on freedom of the seas. As inspection was to be primarily by observation, this could, in their view, already be undertaken by any power under the existing law of the sea, not only in territorial waters, but above the shelf, without any need to obtain permission from the coastal state.

Finally, Japan and one or two other states pointed to the problem of the 'gap': between the end of territorial waters (for three-mile states) and the area where the prohibitions would begin—that is between three and twelve miles from the coast—there would be an area where other states might be able to place nuclear weapons or weapons of mass destruction. Some countries accordingly suggested that all military activities should be barred to the non-coastal state in this area, or that at least the consent of the coastal state should be required for them.

As a result of these criticisms, the United States and the Soviet Union presented in October a revised version of their draft. On verification the treaty was slightly strengthened. It was provided that, if consultation and co-operation had not 'removed the doubts, and there were serious doubts concerning the fulfilment of the obligations' of the treaty, the parties could refer the matter to the Security Council (where the super-powers had a veto). To meet the demand for a guarantee that the treaty would lead on to negotiations for more far-reaching measures, a provision for a review conference after five years was included, during which there could be consideration whether or not it needed to be extended into other fields. The revised draft also removed the special veto provided for the nuclear weapon powers on amendments. Though they of course retained the right not to ratify themselves amendments made to the treaty subsequently, they could no longer prevent these amendments being made among the other parties. The problem of the 'gap' was dealt with by the inclusion of a paragraph to the effect that the undertakings contained in the treaty should also apply between the edge of the territorial sea and twelve miles from the coast 'except that within that zone they should not apply to the coastal state'.

These changes, however, did not satisfy other states, and during the early months of 1970 the two super-powers made a further effort to meet the objections. On April 23, after the Conference of the Committee on Disarmament had reassembled, the United States and the Soviet Union presented yet another draft. This met one or two more of the criticisms made. A bow was made to the demand of the Latin-American countries that the consent of the coastal state was necessary for any verification undertaken: it was laid down that 'parties in the region of the activities and any other party so requesting should be notified of, and may participate in, such consultation and co-operation'. This formula ingeniously avoided the implication that the coastal state held any strict legal rights in the area. And as it was a matter of opinion which countries were 'in the region of' any particular activity, it provided no very burdensome obligation. Similarly, verification procedures were to 'pay due regard to the sovereign and exclusive right of a coastal state with respect to the natural resources of its continental shelves under international law'. This successfully indicated that the only legitimate interest of the coastal state was in respect of the resources of the shelf, rather than of any other activities within the area. Finally, a part at least of the Canadian proposal concerning verification was accepted. There was now provision for *inspection* as well as observation when any activity was challenged, but still only by 'co-operation': it was laid down that 'if doubts persist' (after consultation and co-operation), the two states shall 'co-operate' in such further 'procedures of verification as may be agreed, including appropriate inspection of objects, structures, installations, and other facilities'.

There was still no provision for the direct help of the Secretary-General in verification. Still less was any special international machinery for verification set up, as many delegates had demanded.

Nor was there any direct obligation on more advanced countries to help the less advanced in this respect: verification could be undertaken 'by any state party using its own means, or with the full or partial assistance of any other state party'. Finally, there was no provision in the operative paragraphs of the treaty for a continuation of negotiations on other types of disarmament in the sea-bed, as Sweden had demanded. Indeed, in a sense the provision for a review conference after five years made it less likely that meaningful discussions would take place before that time.

Some of these points were still pressed during the summer of 1970. One or two more changes were accepted. The Swedish proposal that the treaty should contain a specific commitment, in an operative clause, to further negotiations was embodied in a new Article 5. The verification procedures were slightly strengthened: when a dispute occurred all states parties 'concerned' would then 'co-operate in such further measures of verification as may be agreed including appropriate inspection of objects, structures, installations and other facilities that reasonably may be expected to be of the kind banned in the treaty'. A report on the action taken would be made to other parties. Finally, a reference was included to verification if necessary 'through appropriate international procedures within the framework of the UN and in accordance with its Charter'. This final provision was eventually included as a compromise, after the Soviet Union had strongly resisted a Canadian proposal that the treaty should provide for the 'good offices of the Secretary-General' to be invoked in such cases (reflecting general Soviet opposition to according a special role to the Secretary-General).

The treaty, so amended, was generally accepted in the next Assembly, which was not disposed to quibble further. It was approved on December 7, 1970, by 104 votes to 2, with two abstentions. It was signed by 67 countries, including the three main nuclear powers, on February 11, 1971. It entered into force when the necessary ratifications had been received in May, 1972.

Conclusions

Thus, although a start in demilitarizing the sea-bed had been made, the discussions concerning the treaty had shown a number of features, some potentially troublesome.

First, the treaty was to a large extent a super-power creation. All the substantive drafting was undertaken between the United States and the Soviet Union. These then from time to time came down to the multitude to present some new version of the draft they had agreed. The only opportunity that other nations had was in effect to criticize and to exert pressure, not to take part in the drafting itself. Even specific amendments, such as those proposed by Sweden and Canada, were not substantially discussed and voted on within the Committee; they were proposals made to the super-powers, which then went along to consider them and decide to refuse or accept. Some of the objections raised by the other powers—the need for a specific undertaking to negotiate on other aspects of sea-bed disarmament and the recognition of the rights of coastal powers—were ultimately partially met, somewhat reluctantly, in the final version. But the most important, such as the Canadian demand for more watertight verification measures, and the general desire for an international control machinery, or at least the good offices of the Secretary-General, were not met, as these were the points on which the two super-powers themselves also had the strongest views (the United States in the first case, the Soviet Union in the second). In other words, as it was essential to get super-power ratification, they were in practice accorded a kind of veto on every proposal.

The second striking feature of the negotiations was the total reversal of the position of the Soviet Union between August and October, 1969. This was reminiscent of earlier abrupt Soviet concessions on political issues; for example, on Trieste and the Austrian State Treaty in 1954. It was no doubt largely tactical. There is little doubt that the Soviet Union would still have preferred a total ban of the kind it first demanded; but eventually it came to the conclusion that even a partial treaty was better than nothing. The Soviet Union preferred to clinch a deal on this immediately (so achieving a further demonstration of super-power understanding) rather than continuing to make the best the enemy of the good.

More significant than these points, however, in the context of the sea-bed issue as a whole, is the light which negotiations threw on the attitudes of various powers towards certain central sea-bed issues, not all related to disarmament. The almost paranoid concern of the Latin-American states to preserve their own continental shelves and waters (under their own definitions) from unwelcome visitations by other powers is an indication of their general view on other rights—for example, the right of scientific research, the right to control pollution within the continental shelf, and the rights of any enforcement agency there, not to speak of the right of exploitation. The role of such nations as Sweden and Canada, which played an effective part as gadflies to the super-powers and succeeded finally in bringing about a significant modification of their original proposals, possibly foreshadowed the kind of influences that such middle powers could sometimes exert on other issues in future sea-bed discussions.

Equally notable was the solidarity of the super-powers in the face of these various proddings and probings, particularly in view of the diametrically opposed views which each had put forward in the early stages. Because agreement on any aspect of the sea-bed is likely to be at least partially dependent on the goodwill of both powers, their capacity to make common cause in this way is perhaps in some respects to be welcomed. But as also, on some of the broader issues of the sea-bed, the views they hold are not always among the most progressive, it is perhaps to be hoped that on other matters the remaining powers will be somewhat more successful in securing changes of view among the super-powers.

When the treaty was finally signed, many greeted it with derision. It was widely mocked as a solemn undertaking, entered into by many powers, not to do something which they never had the slightest intention of doing anyway. This was in fact a quite unwarranted judgement, which could only have been made by those who were ignorant of the facts. As we saw in Chapter 3, the statements of many military spokesmen and writers had given good reason to believe that nuclear powers might see advantage in the emplacement of nuclear weapons on the sea-bed to escape detection or to reduce distance. And it was certainly the possibility of nuclear uses of the sea-bed that had aroused the most concern. If that danger was genuinely removed or lessened by the treaty, it was not an achievement to be despised.

The real criticism is quite different. The fact is that the treaty does not even prohibit all uses of *nuclear* weapons from the sea-bed. It specifically does not include weapons fired from submarines or other vessels resting on the sea-bed: at an early stage after the introduction of the US–Soviet text on October 7, 1969, the U.S. delegate made unmistakably clear that, as the treaty was concerned with the uses of the sea-bed and not of vessels on the sea-bed, a submarine anchored to, or resting on the sea-bed would not be violating the treaty. And to make this doubly clear, the treaty explicitly refers only to 'fixed' installations. So a barge or other vessel, enclosing a missile, and almost never moved, would not be covered.[1] Nor, of course, would missile emplacements within the territorial sea. Equally important, the treaty has done nothing to reduce the dangers of military uses of many other kinds. All the other possible types of military activity in the sea-bed described in Chapter 3, in some ways more immediate, remain unaffected. The use of missile or torpedo emplacements, whether against shipping or against territory, with conventional warheads, the establishment of manned bases among the various peaks and seamounts of the ocean, the emplacement of mines in the soil of the sea-bed, these and others remain a very real possibility. And as sea-bed technology, including the technique of putting men in the sea at lower and lower depths, advances, so these various possibilities move nearer.

Even in the field which the treaty attempted to cover, its provisions were not all-embracing or without loop-holes. The commitments made were supposed to embrace missiles and installations 'specifically' for use with nuclear weapons. But who was to say which installations were 'specifically' for nuclear weapons or weapons of mass destruction? If the power concerned claimed that they were *mainly* designed for conventional warheads, but could additionally be used for nuclear weapons, were they permissible? Even if such a missile was genuinely designed for conventional purposes, it might be relatively simple, at a time of crisis, to convert it to take a nuclear, instead of a conventional, warhead.

Thus the treaty is riddled with loop-holes available for the unscrupulous to exploit. To ensure more effective demilitarization of the sea-bed, therefore, a new and more comprehensive treaty is still needed. Whether or not this comes about will depend on how vigilantly the members of the Disarmament Committee ensure that the provisions of the treaty relating to renewed negotiations are effectively implemented. So far, five years after the treaty was signed, those further negotiations have not even

[1] The U.S. ULMS (Underwater Long Range Missile System) and SABMIS are designed to establish medium and long-range ballistic missile systems in specially designed submersible vessels which can operate from 400–4,000 metres in depth.

begun. The real test for the international community, if some more effective measure of sea-bed disarmament is to be implemented, will be how vigorously it pursues the attempt to draw up new agreements restricting other types of military activity in the sea-bed.

CHAPTER 7

The Conservation Issue

The Jurisdictional Tangle

But there was another field in which dangers were arising. The possibility of exploitation of the deep seas area brought a new importance to the question of conservation.

The problems of conservation and pollution, as we saw in Chapter 4, had been growing worse for many years. They had been discussed in a number of international bodies long before the sea-bed issue was raised in the Assembly. Each of these was gradually extending its activities in all directions, and there was an increasingly serious conflict of jurisdiction between them.

A new impetus was given to discussion of the subject in 1968. While the report of the *Ad Hoc* Committee on the sea-bed was still being discussed, Sweden introduced an item in the Assembly calling for an international conference on the environment. This was the issue which, in that year, became the main focus of discussion over cocktails in the delegates' lounge: the equivalent in a sense of the Maltese item in the previous year. And this fact perhaps symbolized the eternal relationship between the subjects: if exploitation begins, can pollution be far behind? Certainly, the close historical association of the two items contributed to ensuring that, throughout the sea-bed discussions, the conservation issue was rarely allowed to recede into the background.

In the UN, as in the world outside, consciousness of the problems of pollution, and of the environment generally, emerged only very slowly. Action began, unpublicized, in the fifties. Activity was at first concentrated in very narrow fields: mainly that of the pollution of the sea by oil, and, to a lesser extent, atmospheric pollution. It was only in the second half of the sixties that awareness of the scale of the problem began to be clearly reflected in the direction and content of UN activities. Even then, concern was at first confined to the more developed countries, for whom the problem was most severe. Poor countries suspected a device to distract attention from development problems.

By 1967, when Malta first raised the sea-bed issue, a number of steps were being taken, in a wide range of UN organizations, to improve knowledge of the area or reduce the dangers arising from pollution. There was, however, little co-ordination or order among these various programmes, and continuing conflict among the various organizations concerned. It was clear that a far better system would be required if large-scale exploitation of the sea-bed was to occur.

The Debate on Pollution

There were really two separate, though interrelated, questions. Though both were at first sight relatively straightforward and uncontroversial, both became the subject of increasingly bitter conflict.

One concerned pollution and the means of combating it. The second concerned sea-bed technology and research, and the means of ensuring that effective capacity in these fields was more widely distributed. It is scarcely surprising that in general the former was the main interest of the richer countries, and the latter of the poorer.

Though their attention was primarily focused on the more enchanting prospect of future sea-bed riches, most governments paid at least lip-service to the environmental cause. It was not, in itself, a contentious question. It was universally taken for granted that any régime must provide adequate safeguards against pollution. And the need to provide these safeguards became one of the many pious platitudes which abounded in the Sea-bed Committee's discussions. The real question in dispute was: what were the obligations of governments in this field and how should international control, if any, be organized?

There was growing concern that pollution could result directly from sea-bed exploitation. A single blow-out from an off-shore oil well could pollute vast expanses of ocean, destroy the ecological balance within a large area and affect several countries. Apart from the danger of oil or gas, released directly from the well, there were also the harmful effects from materials stirred up from the sea-bed during mining and dredging, and from waste materials dumped in the ocean. The increasing number of oil rigs had already become a significant hazard to navigation and fishing, while the growing number of tankers presented similar problems. Operations in the deep sea-bed would add to these dangers. It was widely agreed, therefore, that there was need for effective legislation, international as well as national, and perhaps some kind of international inspection to reduce these dangers.

At first most members contented themselves with somewhat woolly statements of general concern, and made few specific proposals for action. Iceland, however, called for a Secretariat study on how to protect living and other resources of the sea-bed against pollution. Iceland's interest in this question was not perhaps altogether disinterested. Its government no doubt was genuinely preoccupied about a danger that could threaten the country's main livelihood. But Iceland had also rightly understood that pollution had become, at least temporarily, a universal bogy, against which governmental action was everywhere demanded. Approval of such governmental action might extend to action taken by governments beyond the existing area of full national jurisdiction. And this in turn could lead to nations taking a more liberal view of attempts by governments to extend their national jurisdiction for other purposes—for example, for the conservation of fishing resources, such as Iceland itself planned.

In other words, the question of pollution offered a heaven-sent justification for the view-point, long held by Iceland and the Latin-American states, that rigid lines of national jurisdiction, first laid down many years ago, were rapidly becoming out of date, and needed to be looked at afresh in the light of changing needs and changing technology, and of the special geographical and economic situation of particular states. To this extent, the issue became part of the general conflict concerning maritime boundaries (see Chapter 9 below). That connection became even closer when, at the beginning of 1970, Canada introduced legislation asserting jurisdiction 100 miles from its coast for environmental purposes. Iceland's resolution thus asked that the study it called for should extend to the examination of 'the circumstances in which measures may be undertaken by states for the protection of the living and other resources of those areas in which pollution detrimental to these resources has occurred or is imminent'.

The report asked for appeared in June, 1970. It pointed out the importance of arriving at a legal definition of pollution. Most human activities in the sea caused some harm to the environment, and if pollution were absolutely prohibited, no exploration and exploitation of mineral resources there could take place. Either therefore, some maximum level of permissible change must be established, which would require detailed scientific enquiries and probably a full-scale system of monitoring; or the matter could be left for legal action between states. National states could probably obtain redress through legal action where specific interests were affected, but this would be far more difficult where the damage was to the international environment as a whole. There might therefore be consideration of some system for providing financial security, with a ceiling on the maximum amount of liability where damage of this kind was caused. This could redress the loss which other users of the sea might otherwise suffer. The report said virtually nothing on the Icelandic point about the right of states to take action beyond territorial waters to protect specific interests, merely noting that claims of this kind which had been made within the Sea-bed Committee had been strenuously resisted by other states.

The desire that there should be some agreed definition of what represented harmful pollution was generally endorsed. A number of proposals were made for action. Iceland wanted the establishment of a marine pollution centre within the UN framework to take the primary responsibility, and a system of ocean stations to monitor the level of pollution at least in areas that were particularly affected. Principles of liability and compensation should be agreed in international treaties, to be implemented both by national governments and UN agencies. Many countries wanted the proposed sea-bed authority to have sweeping powers in the field of pollution control, and to be equipped with a world-wide monitoring or surveillance system and the means to control the policies followed in exploitation and disposal.[1]

More specific proposals about the way in which pollution from

[1] In May, 1971, the Secretary-General presented a further report on marine pollution, proposing the establishment of standards relating to the discharge of wastes into the oceans, whether within or outside national waters. Such standards should be coupled with 'a reporting system' on waste disposal and other forms of pollution. At the same time, since most pollution began in rivers and coastal waters, there was a need for the 'standardisation of national

exploitation should be dealt with were contained in the various draft treaties and other suggestions placed before the Committee from 1970 onwards.[1] The draft treaty presented by the United States (see page 182 below) provided that the proposed International Sea-Bed Resource Authority would have responsibility for the inspection of all licensed activities throughout the sea-bed area beyond 200 metres in depth with its own inspection service. The Authority would apply strict pollution standards in granting licences in the deep-sea areas. If it found that its regulations were being violated, it could demand that steps should be taken to remedy the situation, and, in the final resort, bring the matter to a tribunal to be set up under the treaty. Similarly, any state which believed it had suffered damage from the failure of another state to maintain adequate safeguards could bring a case before the tribunal. The tribunal could impose fines and the payment of damages, and, if necessary, the revocation of the licence. Similarly, the Maltese draft treaty proposed that all activities in the international area were to be conducted 'with strict and adequate safeguards for the protection of human life and for the protection of the marine environment' (Article 74). Even in national waters states were to ensure that 'special precautions are taken in the construction and siting of installations . . . which may cause deleterious effects to the living resources or to the quality of the marine environment'. Nearly all the other draft treaties suggested more or less detailed regulations on this question.

A further advance was made at the Stockholm Conference in June, 1972. A new environmental agency, within the UN, was set up at Nairobi. A $100-million fund to deal with international environmental problems was established and a new Council to control it. The new environment programme (UNEP) would have some responsibility for the seas. It could set up a permanent monitoring system to measure levels of pollution all over the world, and give assistance to governments and regions. The special

---

regulations for the control of pollution', the extension of existing national schemes to regional networks, and finally the integration of all in a global system. International assistance should be given for investigations in tropical and subtropical areas. New international legislation was required, to be the responsibility of the organisations concerned, especially IMCO and later, perhaps, the proposed sea-bed agency.

[1] See p. 182–93 below.

importance of marine pollution was recognized, and 23 principles on that subject were adopted. A subsequent conference in London in 1972 reached agreement on the terms of a new Convention on the dumping of dangerous substances in the oceans.[1] And in 1973 a still wider measure on discharge from ships was agreed in IMCO.

From 1971, the Sea-Bed Committee, in preparing for the proposed Conference on the Law of the Sea (see below, pages 150–2), began to address itself to drafting treaty articles on pollution. It rightly decided that it must consider pollution in territorial waters and the continental shelf as well as in the high seas (because one could affect the other) and would concentrate on legal rules rather than technical regulations, which could follow later. A number of proposals, more or less ambitious, for the creation of new international institutions and new international obligations in this field were made. Draft articles on the subject were put forward by Canada, Malta, Australia, the Soviet Union and other countries. These were generally more comprehensive than the proposed treaties since they covered pollution from all sources: the Soviet Union, for example, wanted *governments* to take responsibility for all pollution caused by their subjects. In any case, the Conference on the Law of the Sea should, it was widely held, formulate general obligations in the field of pollution. Conventions on specialized subjects such as those being prepared by IMCO on shipping, or regional arrangements, should be fitted into this framework.

There was, however, little agreement about the details. The exact form which new obligations of this kind should take still remained to be decided. It was one of the many questions that was supposed to be finally settled at the great Conference on the Law of the Sea that was to open in 1974.

The Debate on Research

This debate on sea-bed pollution was accompanied by another, somewhat loosely related, on technology and research. This one was rather more acrimonious.

A major conflict of views developed here. The wealthier

[1] A more comprehensive agreement for the north-east Atlantic (The Oslo Convention on the Prevention of Marine Pollution by Dumping of Wastes and Other Matters) was reached among Western European nations earlier in the same year.

countries favoured an extensive programme of ocean research, both for purely scientific reasons and because it could promote the development of knowledge useful for exploitation and pollution control. The developing countries had quite a different interest. They were not, in general, very concerned about pollution. They certainly did not want UN research funds to be used to assist the national exploration programmes of the richer countries. They did not want research undertaken by others in their own continental shelves. What they did, desperately, want were UN programmes which would help to redress, however inadequately, the technical imbalance between the rich countries and the poor.

This technological gap had become as important to the developing countries as the income gap. For it determined their capacity to reduce the latter, and to share in technological advances of many kinds: in electronics, computers, space, telecommunications, earth satellites and much else. It was, in particular, a problem fundamental to the whole sea-bed issue. For so long as they did not possess the necessary technology, they were likely to be excluded from any active participation in exploitation, and so, whatever they might be granted in the way of royalties, prevented from sharing in the *direct* profits from the seas.

The need for greater knowledge was not in dispute. The *Ad Hoc* Committee in 1968 referred to the need for 'concerted exploration in its broadest sense . . . to gain knowledge and an understanding of the properties of the ocean floor, of the nature of the marine mineral deposits and of their distribution'. There was a suggestion for 'an internationally co-ordinated programme to explore the economic aspects of a few specific areas', and the Committee endorsed the proposal put forward by the United States for an International Decade of Ocean Exploration. The developing countries insisted that both this and any UN programme should concentrate on helping to strengthen the research capabilities of poorer countries. They thought that the UN and regional bodies should be given more responsibility by providing technical assistance in developing continental shelf resources. Without this, they might be unable to develop their own shelves except by giving concessions, perhaps on unfavourable terms, to the big international companies. They thus quoted with approval the example of the UN Economic Commission for Asia and the Far East's Committee of Co-ordination of Joint Prospecting for Off-shore Resources, which had undertaken some work of this kind.

Changes were meanwhile being made in the structure of the international machinery relating to research. An inter-agency board to advise the Intergovernmental Oceanographic Commission (IOC) was proposed, including senior representatives from each of the main agencies concerned with such matters. This was intended to ensure harmonization between the expanded programme of research under the IOC and the various programmes that each agency had already launched independently. The financial resources would come partly from the participating agencies, and partly from a trust fund (that is, a fund established for a specific purpose) for the IOC itself, which was administered by UNESCO. Within the IOC, a new Executive Council was set up to replace the much looser Consultative Council that had existed before. The statutes were revised in a way which strengthened the IOC's links with organizations other than UNESCO. Finally, on the official level an Inter-Secretariat Committee on Scientific Programmes relating to Oceanography (ICSPRO) was established.

In 1968, the IOC was asked by the Assembly to co-ordinate 'a long-term and expanded programme of world-wide exploration of the oceans and their resources', as well as 'international efforts to strengthen the research capabilities of all interested nations, with particular regard to the needs of the developing countries'. Here, therefore, a kind of compromise was reached between the rival objectives. The programme would meet the demand of the rich countries for an expanded scientific programme, nationally financed but internationally co-ordinated, as well as that of the developing countries for efforts to assist poorer countries to improve their research capabilities. Even so, many poor countries felt that the programme as proposed did not adequately respond to their main needs. While it generally endorsed the programme, the Assembly demanded it provide more opportunities for developing countries to participate and improve their research capabilities, for example in regional programmes: whether for scientific study of particular areas, or for participation by countries of a particular region. The demand for greater participation in this programme was probably a somewhat forlorn plea: as each section was financed by a national government, its own nationals were bound to play the major role. More likely to win a response was the hope that *separate* UN or bilateral programmes would be established, to help in training experts from among the developing countries so that eventually they could participate on more equal terms in exploration and exploitation. The long-term programme was revised by the IOC in the light of the various comments, and the General Assembly endorsed it at its 1969 session.[1]

During the following years, the developing countries continued to stress the need for training and assistance in research. This has therefore become one of the main issues in the sea-bed discussions, with the poor countries demanding more specific commitments, and the rich ones generally somewhat evasive. Individual proposals and draft treaties put forward in the committee have suggested specific arrangements for this purpose. The U.S. draft treaty provided that the Council of its proposed Sea-Bed Resources Authority should be empowered to set up or support international or regional centres, 'to promote study and research of the natural resources of the sea-bed, and to train nationals of any Contracting Party in related science and technology of exploration and exploitation, taking into account the special needs of developing states. . . .' The Tanzanian draft treaty provided that its proposed international sea-bed authority would 'encourage and assist research and the development and practical application of scientific techniques for the exploration of the international sea-bed area and the exploitation of its resources and to perform any operation or service useful to such research', including providing 'services, equipment and facilities'; it was also to 'promote and encourage the exchange and training of scientists and experts', and to provide upon request, 'technical assistance and experts for the use of developing countries in oceanographic exploration and exploitation'. The Council of the proposed international sea-bed authority was to be responsible for provision of technical assistance to developing countries 'to increase their capabilities to participate in activities related to the area'. The Maltese and other treaties had similar provisions for the transfer of technology.

Another subject of dispute concerned legal rights to undertake research in the sea-bed off the coast of another state. The 1958 Convention on the Continental Shelf provided for freedom of scientific research within the shelf. The rich countries sought to show that such freedom was therefore a recognized principle of international law. Provided that certain conditions had been complied with, that notification had been given, that information

[1] G.A. Resolution 2560 of 1969.

obtained was published and made available to the coastal state, and that no commercial investigations were undertaken, rich states felt that permission should be virtually automatic and not subject to the discretion of the coastal state. Most poor countries denied this, claiming that there was no established body of law on this point and reserving the right of the coastal state to refuse permission for research, whether commercial or purely scientific, in the continental shelf or economic zone. They held that it might not prove easy to distinguish genuinely scientific activity from commercial investigations, and all must be strictly controlled.[1]

Even in the apparently uncontroversial field of scientific research therefore there were political disagreements. There were differences on the relative importance of pure research on the one hand and the training and assistance on the other. There were conflicts about the rights of coastal states to control research off their coasts. There were differences concerning how far any training that was given should be provided through bilateral programmes, as the Russians, for example, demanded, or through the activities of international agencies or the new sea-bed authority. Above all, there were divisions concerning the *scale* of the assistance to be provided. All were conflicts arising between richer and poorer states.

All of these questions would be the subject of bitter and protracted battles at the great Conference on the Law of the Sea which had been proposed.

Conclusions

Thus the questions of pollution and research, though both were at first sight relatively uncontroversial, in fact have each aroused considerable dispute. In both cases, this controversy partly concerned organizational questions: what institutions would be required to undertake the tasks necessary in each field? But, in both cases, it mainly concerned more fundamental problems: what *were* the tasks that should be performed in each area? And

[1] There was a paradox about this position. The fact that scientific investigation might always lead to economic applications meant there was the *more* reason for coastal states to allow it to take place freely. For economic *applications* could only take place with their knowledge and authority and would bring them substantial benefit, so investigations should logically have been encouraged rather than deterred by them.

in both cases too, the division reflected, more perhaps than on any other sea-bed issue, a clear-cut difference of view and interest between rich countries and poor.

On pollution, the poor countries had at first been only marginally interested. They had regarded the problem as being not merely primarily the concern, but primarily the responsibility, of the rich. They acknowledged that there was a need for strict rules relating to pollution as part of the sea-bed régime. But they felt that compliance with these rules was the responsibility of the enterprises which undertook the exploitation. And they were certainly reluctant to see large expenditures from international sources, especially from the revenues of the régime, to pay for any programmes required for this purpose.

The rich had been aware earlier of the pressing nature of the problem, in part because it inevitably affected them first, and in part because of the wave of intense preoccupation with environmental questions which enveloped these countries at this time. National, and occasionally regional, measures in the field had been undertaken by them for some years. Much of the scientific and industrial knowledge that was required for dealing with the question was concentrated amongst these countries. But they felt that the problem was a global one, which had to be tackled on an international basis if the measures were to be effective. This might require international spending, and would certainly involve action by many poorer countries as well as the rich—for example, on river and coastal pollution.

On research too, there were sharp differences of opinion and emphasis. The poor countries were insistent on the right of the coastal states to exercise control over such activities, at least in the continental shelf and sometimes further from their coasts. The rich tended to stress the need for total freedom of scientific research throughout the sea-bed, including the continental shelf. The poor wanted a strict obligation to publish all results of research in the sea-bed within a short space of time. The rich wanted greater flexibility on this. The poor wanted the long-term research programme to be orientated closely to economic applications; the maximum opportunity for their own nationals to take part in the programme and to be trained in marine skills; and control to be exercised by a primarily political body, such as UNESCO, in which their voices would count for more. The rich wanted the research programme to be mainly pure science, without excessive training obligations and controlled by a largely non-political body, such as they held the IOC to be.

Despite these differences, the main needs began to become clearer as discussion continued. For pollution, it became evident that the first need was to identify the major pollutants and classify their causes. For this there would be need for an extensive scientific survey as soon as possible to establish the present levels of pollution of different kinds: the *base-line* against which future changes would be measured. This would require some agreement on the yardsticks to be used in measuring pollution: for example, the biological species which would be regularly and systematically examined in the different areas of the ocean. It would involve the standardization of techniques and instruments used in different countries for this purpose. It might then be possible to establish a permanent system for monitoring the ocean, beginning perhaps with co-ordination of national and regional measures (with technical assistance for developing regions), leading to an international system, possibly based on the Integrated Global Ocean Station System (IGOSS).

This improvement in knowledge could be accompanied by new international legislation. The IMCO Conference of 1973 has sought to extend the existing regulations relating to shipping, both by tightening the rules relating to the discharge of oil, and by extending them to cover other substances: these are intended to abolish all deliberate discharge of oil within the next few years. New international conventions have been prepared to deal with similar problems, such as the dumping of industrial wastes in the oceans, or at least certain parts of them. National legislation relating to the discharge of wastes in rivers and coastal areas will need to be extended and harmonized, to limit, for example, the amount of sewage, mercury, DDT and other wastes entering the oceans from this source. A register of all dumping activities will need to be established, so that at least a greater knowledge can be obtained of the amounts of different substances being placed in the oceans and the kinds of regulation which might be required. More effective international policing of these various regulations and conventions may need to be instituted, perhaps some kind of world ocean pollution inspectorate.[1] The final outcome may be a combination of national, regional and global measures.

[1] A Marine Environment Protection Committee, with monitoring and policing powers, has been proposed to IMCO by a U.S. spokesman.

The really controversial questions concerned the institutional arrangements to be made and the question of jurisdiction. Some activity might be organized at a regional level, especially for the enclosed and semi-enclosed seas, where pollution was worst: some international arrangements relating to the Baltic, the Mediterranean and the North Sea are already being made. At a world level, the Stockholm Conference has decided in principle on a world monitoring system. Given the powerful constituencies that they serve, existing agencies will in all probability retain their responsibility in the areas in which they already possess expert knowledge and capacities. But for monitoring of ocean pollution a new unit, possibly associated with the IOC, or with the IGOSS, may be required. And there would certainly be a need for some intergovernmental co-ordinating body, to keep the entire field of ocean pollution in its different aspects under review. Possibly a new Oceans Council is now required which might take on overall responsibility in this field (but not for controlling exploitation—see page 223 below).

Some of these problems are being tackled in the context of action on the environment generally. The new UN Council on the Environment will, it is to be hoped, acquire a considerable degree of authority over the agencies. Beneath this there may be intergovernmental bodies in specialized fields, including marine pollution, working closely with the different agencies concerned. There may in addition be regional bodies responsible for research and training, possibly supported by the general Fund for the Environment (as well as by national and individual contributions). Finally, there will certainly continue to be a role for non-governmental scientific bodies, such as the Scientific Committee on Oceanic Research (SCOR), which will be concerned with the scientific aspects of pollution.

Most of this will be in place before any sea-bed authority has been established, which is unlikely to come before 1980 at the earliest. By then its role in this field could have been largely pre-empted. The question that would have to be decided then, therefore, is: how would any sea-bed authority be fitted in with this set-up, and what responsibilities should it have in the pollution field? The probability is that, unless there is a radical change of thought in the meantime, the sea-bed authority will deal only with those aspects of pollution directly associated with sea-bed exploitation. On these aspects it would undoubtedly frame the necessary rules and undertake the necessary inspection. Given that the type of pollution that could result from such exploitation, whether from blow-outs of oil or gas, or from the disturbance of the sea-bed by nodule exploitation, could be even more serious than that which takes place from existing causes, such as the discharge of oil from ships, this in itself will be a substantial responsibility. But it might be better to have a single co-ordinated system for pollution in the whole marine environment. In this case the Environment Programme, or a new oceans council, could exercise over-all authority.

As to research, the Long Term Expanded Programme of Ocean Research will probably remain the main focus. But it should be adjustable to changes which might arise later, including new needs becoming evident from the development of sea-bed exploitation. It might be necessary to give added emphasis to those aspects of the programme relating to pollution: here the two questions of conservation and research become intertwined with each other. Pressures for greater participation by the nationals of developing countries, and for more training in oceanography to be associated with the programme, will undoubtedly continue. It is important that this need should be met as generously as possible. A sea-bed system which provided *revenues* for developing countries but was effectively managed by the nationals of advanced rich countries and their companies would not be international at all in the proper sense of the word. It would certainly arouse increasing resentment among poor nations. Thus the new sea-bed agency will need to devote increasing resources to the transfer of knowledge and techniques to poor countries.

Conservation in all its aspects, therefore, was an important and by no means uncontroversial element of the whole sea-bed issue. Both pollution and research were among the questions which were to be decided by the Conference on the Law of the Sea. If the oceans are to be saved from grave environmental damage, it is important that effective international measures on this subject be agreed by the world community and that they are implemented before the life of the oceans is irreparably damaged.

CHAPTER 8

The Principles

On disarmament and conservation, therefore, progress was slow. But the real issue was more basic: to whom do the resources of the sea-bed belong, and how are they to be exploited? It was on this that discussion increasingly focused.

A significant step forward was made in 1970. The Declaration of Principles[1] concerning the sea-bed was passed almost unanimously in the General Assembly of that year. This Declaration is generally accepted as having laid down the broad principles on which any sea-bed régime to be agreed must be based. It is thus worth looking at how they were agreed and what they say.

The debate on these principles lasted nearly three years. It is not necessary to describe it in any detail. But it may be useful to pinpoint the key issues that were under discussion, as these go to the heart of the sea-bed problem.

The 1968 Assembly had asked the new Sea-bed Committee to 'study the elaboration of legal principles and norms which would promote international cooperation in the use of the seabed'. It was always the intention that the basic principles should be incorporated in a formal Declaration. This would lay down the general framework on which eventually a treaty governing the area would be based.

Some principles—what might be called the platitudes—were relatively quickly accepted: the need for ensuring appropriate safety measures in any exploitation undertaken; the need to prevent 'unreasonable interference' with other uses of the sea; the need to

[1] See p. 139 below.

preserve the freedom of scientific research (even here some of the developing countries, especially the Latin Americans, were dubious, apparently entertaining a lurking fear that armies of innocent-looking oceanologists might be used as a kind of Trojan horse to acquire access to their continental shelves for sinister purposes, economic or military). There was general agreement on the obligation not to create pollution; though there were wide differences concerning the scale of the obligation. It was accepted that the use of the sea-bed should be 'in accordance with international law and the principles of the UN Charter'; but as international law and the Charter both said remarkably little about the sea-bed, this obligation was not very arduous. Again, the principle that any exploitation should avoid interference with other uses of the seas, such as navigation, fishing, scientific research, the laying of cables and so on, appeared relatively uncontroversial; but here too there were differences of emphasis, the Latin-American countries being determined to assert the rights of coastal states to control activities off their shores, and apprehensive that emphasis on the traditional freedoms might be used to resist this right of coastal state control.

There was one important principle which, during the discussions in the *Ad Hoc* Committee, had seemed to be universally accepted: that 'there is an area of the sea-bed which is beyond national jurisdiction'. But even this provoked controversy. It seemed a key point, for the principle ruled out any interpretation of the 1958 Convention which sought to hold that the whole sea-bed, to the mid-point of the ocean or even beyond, could be claimed by coastal states once it had become exploitable. But in fact no government had ever attempted to support this scarcely-to-be-supported contention. Any such interpretation would have been as much against the interests of highly advanced countries, such as the United States, as of developing countries generally. For the extension of national claims to the sea-bed in this way could have excluded the companies of technically advanced countries from large parts of the ocean otherwise free. Developing countries would have lost also, as they could not have exploited the area off their coasts, however large, while the principle would have largely destroyed the case for the sea-bed as the 'common heritage of mankind'; and so for the international sharing of revenues.

But many states, even so, were against accepting this proposed formulation; they declared that it was not a legal principle but a fact (it finally appeared only in the Preamble to the Declaration). The real danger was not that governments would make explicit national claims far out into the sea-bed (though, as we have seen, a few leases had been granted at depths of over a thousand metres). It was that companies would begin operating at greater and greater depths, claiming that because no nation had any rights within such areas, it was open to anybody to pick up what they found there, according to the principle of 'first come, first served'. In this way, the most developed countries would have drawn the entire benefit from the resources. Developing countries therefore wanted a principle accepted that 'no part of this area is subject to any appropriation—either by states or by nationals—by use or occupation, or by any other means': a thesis to which the rich countries would at first by no means commit themselves. Here too, therefore, for long there was deadlock.

The Key Issues

There was a second category of issues on which there were far wider divergences—not of interpretation but of substance.

For example, there was a fairly widespread desire that the area should be used 'exclusively for peaceful purposes'. But even if accepted, this phrase was open to quite contradictory interpretations. Was a submarine resting temporarily on the bottom a peaceful or military use? Was a sonar detector used both for communications and defensive purposes peaceful or military? And so on. Some said that the UN should prohibit all military uses of any kind (even if, by prohibiting anti-submarine devices, it made the waters above become far more dangerous). Others, such as the United States, said that it should prohibit only the *offensive* military use of the sea-bed. This was a very basic difference of view concerning the rules to be established, which is still by no means resolved.

Then, should there be a general sharing of the benefits of sea-bed production? There was a demand for a principle that 'exploitation should be for the benefit of, and in the interests of, all mankind'. This too at first sounded like an important step forward. But in practice all delegations could subscribe to that unimpeachable sentiment, if only because they understood quite different things by it. To the poor countries what it meant was that the revenues from exploitation should be shared on an agreed and joint basis among nations. To some rich countries, however, it meant simply that exploitation should be as much for the benefit of mankind as was all economic activity in any part of the world—in that it would promote economic progress and material development. This did not imply that there should be international control. At the very most, some held, a *voluntary* payment of a proportion of the revenue 'for international community purposes' would fulfil such a principle. They stressed the vital importance of adequate incentives to ensure that exploitation of the area took place as quickly as possible; and would no doubt have argued that the maximum 'benefit and interest' of mankind could be attained by securing the most rapid possible production in the area by the companies which alone had the technology to undertake this. Here was another fundamental difference of attitude.

There were also conflicts on the key proposal for an 'international régime'. The phrase itself was so vague that a wide range of meanings could be attributed to it. To the poor countries, it implied an international authority or control system for running the entire sea-bed, which could ensure effective redistribution of benefits. The Soviet Union held that it meant simply a set of rules governing the area. The principles put forward by Britain proposed that 'there should be agreed, as soon as practicable, an international régime governing the exploitation of the resources of this area': here, at the least, the principle of a régime seemed to be accepted. But not all rich countries accepted even this. The Soviet Union, in particular, demanded that if the hated phrase was to be included at all, it should be in some such phrase as 'a special legal régime shall be worked out'; or, better, that 'legal principles and norms will hereafter be agreed'. Thus there was disagreement not only on the principle of a régime, but even about what the concept meant.

A point still more disputed concerned the proposal for 'international machinery'. This touched one of the rawest nerves of the Communist countries, with their doctrinal aversion to anything smacking even remotely of supranationalism. The developing countries referred to the need to secure the most appropriate and equitable distribution of benefits 'through a suitable international machinery'. The Russians, and their East European allies, refused absolutely to accept such a formulation. As the phrase in itself had little meaning—'machinery' might mean anything from a supranational agency to a consultative committee—this obduracy was perhaps somewhat unnecessary except for symbolic purposes. But it was nevertheless unshakeable. And it was this point, perhaps more than any other, which finally made it impossible to secure unanimous agreement on the principles finally adopted.

But perhaps the most bitter and protracted dispute was on the most central point of all. From the very beginning, Malta had called insistently for the adoption of the principle that the non-national sea-bed area (whatever it was) should be accepted as the 'common heritage of mankind'. In this, she was supported by most of the other developing countries. This was not just a pedantic conceit: the phrase was intended to emphasize and firmly establish, through a general declaration, that the resources of the sea-bed did not belong either to any single set of coastal nations, nor to anybody who found them, but to the world as a whole; and must be used in the interest of all, and with the consent of all. For long the adoption of this principle was resisted by the developed countries, western and eastern alike. They found many esoteric reasons for opposing it. They claimed that it was 'contrary to existing norms and principles'. It was 'devoid of legal content'. And, perhaps more reasonably, that it was open to many different interpretations.

Such arguments made little impression on the poor countries. They continued to insist on the importance of this principle. Eventually, the rich countries apparently decided to adopt the policy: if you can't beat them, join them. One by one, including the Russians, they began to decide that they could accept the phrase after all; but they were careful at the same time to place their own reinterpretation upon it. Their position was only gradually eroded. All the Commonwealth countries, as the result of some skilful lobbying by Malta and a number of British bishops, had been induced to take this view as early as January, 1968, at the Commonwealth Prime Ministers' Conference in London. Other developed countries began to change their views too. Gradually, the rest were picked off one by one. Even the Soviet Union finally shifted its position, though only by dint of declaring that the phrase had no legal meaning. Since the developed countries had all sought to suggest from the start that the phrase was without legal content, there was perhaps little reason why they should have been so opposed to accepting it from the beginning. Conversely, if they had such strong objections to it in the early days, it was hard to see how they could now insist that it was without meaning.

Such verbal disputes, though they may seem arid and profitless, in fact always conceal real and concrete problems, on which the interests of states differ radically. The debate thus serves to illustrate some of the main concerns of the principal countries and groups engaged in discussing the sea-bed issue. Disputes over single commas or definite articles may, in the UN system, sometimes conceal the profoundly and passionately felt objectives of individual nations.

The Debate

It is not necessary to recount here the course of the debate that took place on individual issues. Prolonged battles took place on many of them.

A more fundamental difference concerned the scope of the principles. The rich countries mainly claimed that the principles should be brief and in very general terms, to cover basic points which everybody accepted. To be valuable, they must be agreed unanimously. But this meant that they would also be a minimum set, amounting to very little. The developing countries favoured a far more explicit and comprehensive set of principles, setting out clearly obligations in a number of different fields. This inevitably meant that they brought out points on which considerable disagreement existed. In time, if only because of the pressures of numbers on the side of the developing countries, the scope of the draft Declaration did gradually extend. From the seven brief sentences originally proposed by Britain, the declaration of principles grew to fifteen separate paragraphs, some of them containing a number of sub-clauses.

On some of the more uncontroversial points, the Committee moved towards agreement. It was accepted that the international area (whatever it was) was 'not subject to national appropriation' and that no state could 'exercise or claim over any part of the area sovereignty or sovereign rights' (but we noted earlier the varying reasons why all were ready to accept this); that the area should be 'reserved exclusively for peaceful purposes' (but no definition was provided); that any régime established should provide for the 'rational development and equitable management of the area' (but what was 'rational' or 'equitable'?); and that there should be 'reasonable regard for the interests of other states in their exercise of the freedom of the high seas' (but what was 'reasonable'?). These were of course among the pious platitudes to which all could subscribe relatively easily, but which were either so vague that they concealed a good deal of disagreement over the substance, or else amounted to nothing very much anyway. On the central issues, concerning the nature of the system to be applied, there remained wide disagreements.

On some points positions began to move closer by 1970. There was a hope that a Declaration of Principles could be prepared, to be issued at the anniversary session of the UN in the autumn of that year. Protracted debates and informal consultations during that year, however, failed to bridge the gap. Informal consultations were held in June in New York and later, at the end of July, in Geneva, to produce agreement. But when the 1970 session of the General Assembly opened, the hope that a Declaration might be agreed and passed at the anniversary session still seemed remote. More informal discussions took place, in considerable haste, in the corridors of the Assembly. Eventually, most of the points of difference were narrowed down. And, after a final scurry of consultations, though not in time for the twenty-fifth anniversary itself (October 24), the last gap was bridged. Agreement was reached in November on a general Declaration of Principles within the Committee. Only the Communist countries and a handful of others felt unable to give their full support to this; and even these were content with an abstention. The Declaration was carried by a vote of the General Assembly on December 17, 1970, by 108 votes in favour, none against, and fourteen abstentions.

As always, all the main groups had had to give something away. The developed countries had probably to give most, partly because they were very much in a minority. The Declaration was far more detailed and comprehensive than most of them at first hoped. They had to accept the statement, in the very first principle, that the sea-bed and its resources 'were the common heritage of mankind'. They accepted that all exploration and exploitation should be 'governed by the international régime to be established'; and even though they still maintained that the régime itself would have to be agreed among states, this made it considerably more difficult for them to initiate, or even to condone, activities by their own companies within the sea-bed which were not in accordance with an agreed régime. The régime was to 'ensure the equitable sharing by states of the benefits derived' from the area, and was to provide 'expanding opportunities in the use' of the area for all states. States were to have the responsibility for ensuring that all activities in the area, 'whether undertaken by governmental agencies, or non-governmental entities or persons under its jurisdiction' should be carried out in conformity with the international régime to be established. They conceded the right of coastal states 'in the region of' economic activities in the area to be consulted about such activities 'with a view to avoiding infringement of' the 'rights and legitimate interests' of such states. Finally, they even accepted that the régime to be established would include 'appropriate international machinery to give effect to its provisions'—the phrase that perhaps more than any other, with its implication of a supranational authority, forfeited the support of the Communist states.

The developing countries also had to accept concessions. They had to concede the important point that the international régime should be 'established by an international treaty of a universal character, generally agreed upon'; so acknowledging that the régime could not be imposed by the Assembly by a majority vote but must be adopted with the consent of all those who were to be bound by it. They had to agree to the omission of any statement, such as that they had originally proposed, that 'the United Nations . . . shall take adequate measures to assure the observance of these general principles . . .', which had implied a policing role for the UN. The reference to the 'equitable sharing by states' of the benefits derived from sea-bed exploitation, which could be interpreted in almost any way desired, was considerably vaguer than most of the developing countries themselves would have wished. The Latin-American countries had to give something away on the rights of coastal states: the principles declared on this point gave such states no new rights, but merely demanded that the exploiting countries should pay 'due regard to' existing rights and 'legitimate interests' of coastal states—whatever they were—and consult them concerning those rights and interests.

Thus the Declaration was inevitably a compromise, with all a compromise's strengths and weaknesses. It reflected the lowest common multiple of agreement on the basic issues concerning the area. But it was not only a lowest common multiple. Compared with what most of the developed countries were originally prepared to accept, the statement had moved a very long way forward. The great majority of the states of the world, certainly all those who had voted in favour, had been brought to accept, if only verbally, commitments which only a few years ago many would have regarded as bizarre or even totally unacceptable—essentially, a commitment to regard sea-bed resources as the world's common property, to be commonly controlled.

Inevitably some of the principles remained open to varying interpretations. But on some of the fundamental questions there could be little disagreement. Exploitation was to be according to an agreed régime, and not on the basis of 'first come, first served', or of extended national claims. Exploitation was to benefit all countries. International machinery was to be established to ensure that all governments and companies under their jurisdiction abided by the régime to be established. Last but not least, the sea-bed would not be subject to appropriation, whether by individuals or states, or to any form of sovereignty.

Thus there were at least some unambiguous and positive results from the years of strenuous negotiations which had gone into the formulation of principles.

Conclusions

So, at the end of 1970, three years after the sea-bed issue had been first raised at the Assembly, a small step forward had been taken: an agreed set of principles governing the use of the sea-bed had been almost unanimously endorsed by the UN. How significant was this progress?

It is easy to make fun of the 'principles' that were enunciated. As we have seen, sometimes agreement was reached only by the simple device of re-interpreting particular phrases to the convenience of each participant. Those who had strongly opposed the idea of the common heritage of mankind were eventually prevailed on to accept it, but only by declaring that it had no real significance or legal content. All were able to agree that the area should be used 'exclusively for peaceful purposes'; but they had quite different ideas on what this meant. All could accept that the area should be used 'in accordance with the régime to be established', but still had totally conflicting views on what that régime should be. They agreed that exploitation should be carried out 'for the benefit of mankind as a whole', but differed on what this meant and how it should be achieved. According to this sceptical view, the principles merely represent the points on which there had been a measure of agreement from the start—which meant those that were unimportant. On the significant points, it might hold, agreement was no nearer at all.

Though there is obviously an element of truth in this contention, it was not the whole story. Agreement had been brought nearer even on some points of substance. When the question was raised only three years earlier, there really did seem a possibility that nations might seek to extend their claims to exploiting rights indefinitely into the ocean. This at least was a fear which had been widely expressed, and to which some public writings and statements lent some credence. The universal acknowledgement, incorporated in the preamble to the principles, that 'there was an area beyond the limits of national jurisdiction', was therefore of some importance. It represented a genuine change in the previously existing situation. And it was a declaration which, in practice if not in theory, inevitably to some extent committed even countries which had not subscribed to it, as it would be politically almost impossible for any country (China, for example) at a later stage to seek to assert publicly unilateral claims of the kind which the Declaration explicitly rejected.

Even the reference to the use of the sea-bed only for peaceful purposes, however vaguely worded, represented a commitment of a significant kind. If, for example, in the future, defence departments began to play with the idea of establishing fixed installations for offensive weapons, even of a conventional kind, on the sea-bed, they would undoubtedly be strongly discouraged by their foreign office colleagues, on the grounds that such an act, if discovered, would be universally condemned by world opinion as a violation of the principles adopted. In other words, the political inhibitions to the militarization of the sea-bed were significantly strengthened.

But more significant than these was the categorical assertion that 'all activities regarding the exploration and exploitation of the resources of the area . . . shall be governed by the international régime to be established'. For though the nature of that régime was still unknown and subject to dispute, it was at least publicly recognized that there was to exist some kind of international regulation. It would be difficult for any government to justify a unilateral attempt to claim prospecting rights which were independent of an agreed system of exploitation, or even to fail to take action against companies which asserted such rights, given the clear commitments made under this principle. In short, the situation of anarchy in the deep sea-bed that had been previously widely acknowledged and generally accepted, was now formally rejected. The international consensus had changed. Some form of international regulation was, by general agreement, to be established. While the details of that system remained to be decided, it was an advance that there was public and near universal acknowledgement that such a system should be created at all.

Even the general acceptance of the somewhat amorphous conception of the common heritage of mankind was not wholly without significance. Like the acceptance of a régime, and even more forcefully, the phrase emphasized that the resources of the area were generally regarded as common property, to be taken and used only with the consent of all and according to a system agreed by all. In so far as any property rights existed, the phrase implied, these were vested in the world community and could not be simply appropriated through finding and exploiting. Just as many governments assert property rights in natural resources within their territories, the international community now, this phrase implied, was taking the first steps in asserting its property rights in sea-bed resources within the international area.

However, perhaps the most important gain achieved through the endorsement of the principles was a more intangible one. Quite apart from the significance of individual principles, the fact that a Declaration on the sea-bed had been passed at all was important. For this formally registered the responsibility of the UN for defining the rules to be adopted in the use and exploitation of the sea-bed in the future. Even if there continued to be differences among governments on particular points, it was accepted that it was for the international community as a whole to determine the nature of that régime and how the area was to be used. This was a fairly revolutionary development. There was no other region of the world for which the international community had taken a similar responsibility. For centuries governments had dealt with one another individually in agreeing on the disposition of territories, sometimes even those remote from their own natural borders. Over the Antarctic, the UN had generally endorsed the Treaty under which territorial claims were frozen, and the peaceful use of the area established under mutual inspection.

But this was a Treaty reached elsewhere, among only a few of its members. Over outer space, the UN had negotiated general agreements governing peaceful use and legal obligations; but in that case there were no exploitable resources, or any immediate repercussions for most states from the agreement which was reached. To decide at least on the principle of joint regulation of a large portion of the earth's surface was, in its own way, quite a giant leap for man.

The degree of its importance would of course depend on the agreements to be reached at a later stage: that is, how the Declaration was implemented in the Treaty finally to be drawn up for the area. Among the points still left open, the most vital of all concerned the area that was to be covered by any régime established. It was all very well to agree that there should be a system of joint regulation. But this meant very little without knowing the area it was to cover. Its significance would be small if the regulation accepted only concerned the central regions of the ocean, relatively poor in resources, and left all the main exploitable parts to national states. The first point that had to be determined, if the principles were to be given any content, was: how far do national rights extend and where therefore does the international zone begin? And it is to this point that we must now turn.

DECLARATION OF PRINCIPLES GOVERNING THE SEA-BED AND THE OCEAN FLOOR, AND THE SUBSOIL THEREOF, BEYOND THE LIMITS OF NATIONAL JURISDICTION, 17 DECEMBER, 1970

VOTE: *108 in favour, none against, with 14 abstentions.*

Text of Resolution

The General Assembly,

Recalling its resolutions 2340 (XXII) of 18 December 1967, 2467 (XXIII) of 21 December 1968 and 2574 (XXIV) of 15 December 1969, concerning the area to which the title of the item refers,

Affirming that there is an area of the sea-bed and the ocean floor, and the subsoil thereof, beyond the limits of national jurisdiction, the precise limits of which are yet to be determined,

Recognizing that the existing legal régime of the high seas does not provide substantive rules for regulating the exploration of the aforesaid area and the exploitation of its resources,

Convinced that the area shall be reserved exclusively for peaceful purposes and that the exploration of the area and the exploitation of its resources shall be carried out for the benefit of mankind as a whole,

Believing it essential that an international régime applying to the area and its resources and including appropriate international machinery should be established as soon as possible,

Bearing in mind that the development and use of the area and its resources shall be undertaken in such a manner as to foster healthy development of the world economy and balanced growth of international trade, and to minimize any adverse economic effects caused by fluctuation of prices of raw materials resulting from such activities,

Solemnly declares that:

1. The sea-bed and ocean floor, and the subsoil thereof, beyond the limits of national jurisdiction (hereinafter referred to as the area), as well as the resources of the area, are the common heritage of mankind.

2. The area shall not be subject to appropriation by any means by States or persons, natural or juridical, and no State shall claim or exercise sovereignty or sovereign rights over any part thereof.

3. No State or person, natural or juridical, shall claim, exercise or acquire rights with respect to the area or its resources incompatible with the international régime to be established and the principles of this Declaration.

4. All activities regarding the exploration and exploitation of the resources of the area and other related activities shall be governed by the international régime to be established.

5. The area shall be open to use exclusively for peaceful purposes by all States whether coastal or land-locked, without discrimination, in accordance with the international régime to be established.

6. States shall act in the area in accordance with the applicable principles and rules of international law including the Charter of the United Nations and the Declaration on Principles of International Law concerning Friendly Relations and Co-operation among States in accordance with the Charter of the United Nations, adopted by the General Assembly on 24 October 1970,[1] in the interests of maintaining international peace and security and promoting international co-operation and mutual understanding.

7. The exploration of the area and the exploitation of its resources shall be carried out for the benefit of mankind as a whole, irrespective of the geographical location of States, whether land-locked or coastal, and taking into particular consideration the interests and needs of the developing countries.

8. The area shall be reserved exclusively for peaceful purposes, without any prejudice to any measures which have been or may be agreed upon in the context of international negotiations undertaken in the field of disarmament and which may be applicable to a broader area. One or more international agreements shall be concluded as soon as possible in order to implement effectively this principle and to constitute a step towards the exclusion of the sea-bed, the ocean floor and the subsoil thereof from the arms race.

9. On the basis of the principles of this Declaration, an inter-

[1] Resolution 2625 (XXV).

national régime applying to the area and its resources and including appropriate international machinery to give effect to its provisions shall be established by an international treaty of a universal character, generally agreed upon. The régime shall, *inter alia*, provide for the orderly and safe development and rational management of the area and its resources and for expanding opportunities in the use thereof and ensure the equitable sharing by States in the benefits derived therefrom, taking into particular consideration the interests and needs of the developing countries, whether land-locked or coastal.

10. States shall promote international co-operation in scientific research exclusively for peaceful purposes:

(a) By participation in international programmes and by encouraging co-operation in scientific research by personnel of different countries;

(b) Through effective publication of research programmes and dissemination of the results of research through international channels;

(c) By co-operation in measures to strengthen research capabilities of developing countries, including the participation of their nationals in research programmes.

No such activity shall form the legal basis for any claims with respect to any part of the area or its resources.

11. With respect to activities in the area and acting in conformity with the international régime to be established, States shall take appropriate measures for and shall co-operate in the adoption and implementation of international rules, standards and procedures for, *inter alia*:

(a) Prevention of pollution and contamination, and other hazards of marine environment, including the coastline, and of interference with the ecological balance of the marine environment;

(b) Protection and conservation of the natural resources of the area and prevention of damage to the flora and fauna of the marine environment.

12. In their activities in the area, including those relating to its resources, States shall pay due regard to the rights and legitimate interests of coastal States in the region of such activities, as well as of all other States which may be affected by such activities. Consultations shall be maintained with the coastal States concerned with respect to activities relating to the explora-

tion of the area and the exploitation of its resources with a view to avoiding infringement of such rights and interests.

13. Nothing herein shall affect:

(a) The legal status of the waters superjacent to the area or that of the air space above those waters;

(b) The rights of coastal States with respect to measures to prevent, mitigate or eliminate grave and imminent danger to their coastline or related interests from pollution or threat thereof resulting from, or from other hazardous occurrences caused by, any activities in the area, subject to the international régime to be established.

14. Every State shall have the responsibility to ensure that activities in the area, including those relating to its resources, whether undertaken by governmental agencies, or non-governmental entities or persons under its jurisdiction, or acting on its behalf, shall be carried out in conformity with the international régime to be established. The same responsibility applies to international organizations and their members for activities undertaken by such organizations or on their behalf. Damage caused by such activities shall entail liability.

15. The parties to any dispute relating to activities in the area and its resources shall resolve such dispute by the measures mentioned in Article 33 of the Charter of the United Nations and such procedures for settling disputes as may be agreed upon in the international régime to be established.

CHAPTER 9

The Limits

The Battle Lines

From the time when the sea-bed discussions began, the importance of defining the outer boundary of the continental shelf—and so the beginning of the international sea-bed—has been widely recognized. Dr Pardo, in first raising the question in 1967, pointed to the dangers and uncertainties that resulted from the ambiguous provisions of the 1958 Geneva Convention on the Continental Shelf. Some governments had already been granting licences for exploration in areas far beyond the continental shelf as traditionally understood, certainly well beyond the 200-metre isobath. With the rapid advance in sea-bed technology, these more distant areas would shortly be actually exploited. This could make possible still further claims. It was thus essential to establish a new, and clearer, definition as soon as possible.

For this reason, while the Maltese item was still being discussed in the Assembly, Sweden proposed a moratorium on claims, to counter the danger of creeping jurisdiction. The proposal received considerable support. But demands of this kind, whether for a new international agreement on limits, as Malta desired, or for a moratorium, as Sweden proposed, came up always against the steadfast resistance of some states, and of the Latin-American countries in particular. These countries insisted on the right of every nation to lay down its own sea boundaries. They maintained that the wide limits which they had already declared were firm and irrevocable. Their claims were based on the natural right of any state to the resources of its immediate environment and on the relationship, geographical, economic and social, which existed, they maintained, between 'man, the land and the sea'—a somewhat nebulous concept which nonetheless exercised a profound appeal to some delegates. The 1958 Convention on the Continental Shelf was irrelevant, they held, since relatively few countries had ratified it. Others need not demand 200 miles for this or any other purpose. But it was, as it had always been, for every nation or for every region to decide its own limits in accordance with its own local circumstances. These countries therefore firmly resisted any call, such as that made by Malta, for a new attempt to lay down an internationally agreed boundary, whether for the surface of the sea or its bottom.

The attitude which the Latin-American countries took up on this point was not entirely logical. It is by no means certain, as we shall see, that they had correctly calculated their own national interests in the matter. But in view of the major role which they played, throughout the discussions, in upholding their own claims and resisting or postponing any attempt at an international agreement on limits, it is necessary at this point at least to try to explain it.

In speaking here of 'the Latin Americans', we are guilty of a certain over-simplification. There were some Latin-American countries—for example, Mexico, Venezuela and Colombia—which took up a quite different and entirely independent attitude on the question, by no means supporting the immoderate claims of their neighbours. The leading wide-limit countries were the Pacific coastal states, Ecuador, Peru, Chile on the one hand, and, to a lesser extent, Argentina, Brazil, and Uruguay on the Atlantic coast.[1] These states were fortunate enough to possess particularly valuable natural resources in the waters near to their coasts. In the Pacific, the peculiar character of the up-welling currents, bringing nutrients from the deeper waters towards the surface, meant that the waters a hundred or more miles from the shore were abundant with fish (above all anchovy) in a way known in very few other areas of the world; that was why Peru was the largest fishing country of the world. Their concern was therefore not so much with the sea-bed (for there was virtually no continental shelf in this area) but with the waters above it. And the main aim behind the extension of boundaries was the desire to protect fishing

[1] By 1973 there were ten Latin-American countries which had claimed 200 miles for some or all purposes: El Salvador, Ecuador, Peru, Chile, Argentina, Uruguay, Brazil, Nicaragua, Panama and Honduras.

resources from the attentions of the powerful and highly mechanized U.S. fishing fleet.

Thus the motive of these countries is clear. The argument they used in support of those claims was equally simple, and not without force. For they held that, in extending unilateral claims to protect their economic interests, they were only following an example set earlier by the developed countries. The doctrine of the continental shelf had itself been merely an invention of President Truman in 1945, and one designed by the U.S. government to promote U.S. interests. The United States had found that it possessed important mineral resources within its continental shelf, and had therefore asserted rights of exploitation there to reserve them to itself—a predatory example which other developed countries had been only too willing to follow. They themselves were not concerned with oil or gas, and in many cases had little continental shelf anyway. Such a move, therefore, would be of little interest to them. What were of intense importance to them were the fishing resources close to their coasts. These, they felt, they were quite as justified in reserving, even by unilateral claims, as the United States had been in reserving mineral resources in the shelf to a considerable distance by a unilateral claim in 1945. The basic principle at issue in both cases, they held, was the right of a coastal state to the resources in its immediate vicinity, on which some of its people (and in Peru's and Iceland's case almost the entire economy) had become dependent. This right they regarded as natural and incontestable. And the need to defend it had become particularly urgent because of technological changes in the modern world—above all, the capacity of fishing fleets to travel quickly to distant waters. For this made it possible for the highly mechanized U.S. fleets and factory ships to bear down on their own waters and carry off resources which, they held, rightfully belonged to them. 'Freedom of the seas' in this view merely meant the freedom of rich countries to exploit resources more naturally belonging to poor countries.

Because they were aware that these claims were resisted, and strenuously denied, by other countries, including the United States and the Soviet Union, some Latin-American countries had become deeply suspicious even of any international discussion of the question. They thus sought to prevent, or at least to delay as long as possible, any attempt to arrive at an international definition of the territorial sea and the continental shelf. This opposition became particularly strong as it became apparent, in the late sixties, that there was growing agreement on a twelve-mile limit for territorial waters. For this meant that a new conference, instead of breaking up in confusion as in 1958 and 1960, would probably come out with a large majority for twelve miles, leaving the 200-milers in considerable isolation.

When the permanent Sea-bed Committee met in March, 1969, Malta proposed a resolution on the limits of the continental shelf. It proposed that the Assembly would, by a solemn declaration, confirm that 'the sea-bed and ocean floor and the subsoil thereof, sub-jacent to the water further than . . . nautical miles from the nearest coast, and more than 200 metres deep disregarding rocks and islands without permanent settled population, unquestionably are and must remain beyond national jurisdiction'. The resolution went on to request the Secretary-General to consult with member states and various organizations about the 'convening of a conference at the earliest practical date'. This Conference would revise the Convention on the Continental Shelf of 1958 and formulate legal rules on the exploitation of the sea-bed. The blank would be filled by a suitable number. The Assembly would thus first make a declaration, firmly setting a limit to national claims; and subsequently procedures for revising the 1958 Convention by a conference would be put in motion.

There was little immediate reaction to this suggestion. At the 1969 Assembly, Malta revived the proposal, now in a somewhat different form. The resolution now merely asked the Secretary-General 'to ascertain the views of member States' on the desirability of convening a conference at an early date. This would be for the purpose of 'arriving at a clear, precise and internationally acceptable definition of the area of the sea-bed . . . beyond national jurisdiction'.

At last it seemed possible that a definite step forward might be made in agreeing a sea-bed boundary and so limiting further national claims to the area.

The Call for a Conference

But it was not to be so easy.

The proposal for consultations about a conference went through. But in the process the suggested agenda was enormously expanded. On the proposal of the Latin Americans, that agenda was extended to cover, not only the limit of the continental shelf, but also 'the régime of the high seas, the continental shelf, territorial sea and contiguous zone, fishing and conservation of the living resources of the high seas . . .' and so on. In other words, the agenda would include many highly contentious problems of sea law, some of which many thought had already been decided at Geneva in 1958. This would make the conference less popular and, if it took place, far less likely to end in agreement.

But there was also a danger that the conference would come too late. For the proliferation of claims to ever-increasing areas of shelf continued as before, especially in Latin America. Almost every year saw a new country added to those which claimed a 200-mile limit for some or all purposes. Nicaragua had adopted it in 1965, Argentina in 1966, Panama in 1967, Uruguay in 1969, Brazil in 1970. The habit became infectious and extended even to Africa and Asia. Guinea claimed 130 miles of territorial sea in 1964, Cameroon claimed 30 in 1967 and Gabon 25 in 1970. Ghana established a fisheries 'conservation zone' of 100 miles in 1963, and Ceylon did the same in 1966. Senegal established an exclusive fishery zone of 18 miles and a 200-mile territorial sea in 1971, while Dahomey claimed a continental shelf of 100 miles in 1968.

To check this process there were new demands for a moratorium on claims. A suggestion by Cyprus that 'all states should refrain from claiming or exercising jurisdiction over any part of the sea-bed . . . beyond a depth of 200 metres, and beyond the limits of national jurisdiction they at present exercise . . . pending the clarification of the extent of national jurisdiction . . .' came up against the rooted hostility of the Latin Americans. Instead, a different kind of moratorium was passed: on *exploitation*. This resolution 'declared' that, pending the establishment of a régime, 'states and persons, physical or juridical, are bound to refrain from all activities of exploitation of the resources of the area of the sea-bed . . . beyond the limits of national jurisdiction'; and that 'no claim to any part of that area or its resources shall be recognized'. But the resolution proposed no definition for the international 'area' concerned. It thus could not seriously deter exploitation. On the contrary, it could amount almost to an inducement to governments to extend their claims, to ensure that any economic activities within the sea-bed their nationals undertook should not be covered by its provisions. Moreover, the rich countries mostly stated that they would not be bound by it unless it was accompanied by a similar moratorium on limits. Western states said that they would seek to ensure that their enterprises abided by any régime ultimately agreed, but would not stop them exploiting meanwhile. Nonetheless, this exploitation moratorium was passed by a considerable majority in December, 1969, with voting largely on a rich–poor basis and the rich countries almost universally against.

The Secretary-General proceeded to ask member governments for their views about the proposed conference. The answers he received provided a wide range of ingenious rationalizations in favour of the procedure which would best promote the interests of the government replying. As the existing chaotic state of the law—or non-law—was obviously indefensible, virtually no government came out altogether against a conference. But the Latin Americans tended to argue that there was no value in holding a conference until *after* the régime for the sea-bed was already agreed—that is, for many years. They also still demanded that any such conference should be designed to consider the law of the sea in general, including the high seas, the contiguous zone, fishing rights and the lot, and not just the shelf.[1] Most of the developed countries held strongly that it was pointless to reopen once again all these difficult issues discussed in Geneva twelve years before. They were concerned especially that the conference should concentrate on the two main points that needed to be solved: the limits of the continental shelf and the breadth of the territorial sea—precisely the two questions which the Latin Americans least wanted discussed.

[1] To strengthen their own position and unite their ranks, some of the Latin-American countries, at the instigation of Peru and other fishing countries, organized two conferences during 1970: at Montevideo in May—and at Lima in August. The latter, in particular, came out with a strong statement, somewhat grandiloquently entitled 'The Declaration of the Latin American States on the Law of the Sea' (though it was not in fact signed by all such states). It declared that 'it would be premature to establish the extra-jurisdictional limits of the sea-bed . . . until the above-mentioned stages had been completed'. And it recommended that the participating governments state that they approved of a Conference on the Law of the Sea only if it were to consider all the various subjects proposed for it, including those decided in the Geneva Conventions and only after a 'permanent international régime and the administrative machinery applicable to the extra-jurisdictional sea-bed have been defined, and the studies, reports and enquiries made for that purpose have indicated that there are reasonable hopes for the succes of the conference'.

At the 1970 Assembly, the proposal for a conference was discussed in the light of the answers to the Secretary-General's enquiries. Since, apart from a few hard-line Latin Americans, such as Peru and Ecuador, almost all the answers favoured a conference on some basis, the question was really only how soon and on what terms. There was bitter conflict: on the scope of the conference; on the agenda; and on the date. The United States wanted a conference fairly soon and with a restricted agenda. Brazil wanted it later, with a very broad agenda and a decision on limits only 'in the light of a decision' on the régime. Both the date for the conference and its subject matter were successively changed. Finally, after intensive behind-the-scenes negotiations, Canada, on behalf of a large group of countries, both developing and developed, proposed a compromise which ultimately proved successful. The conference was to take place in 1973 'if possible'. It was to 'deal with the establishment of an equitable international régime . . . a precise definition of the area and a broad range of related issues, including' the various law-of-the-sea issues.[1] The Sea-bed Committee was instructed to prepare draft treaty articles both on the régime and on the various legal issues. The entire proposal to call the conference and to ask the Sea-bed Committee to undertake the preparatory work was then passed by the Assembly by an overwhelming majority.[2]

The resolution marked an important step forward in the discussion of the sea-bed issue. A definite decision had been

[1] The exact terms of reference for the Conference are as follows: 'deal with the establishment of an equitable international régime—including an international machinery—for the area and the resources of the sea-bed and the ocean floor and the subsoil thereof beyond the limits of national jurisdiction, a precise definition of the area, and a broad range of related issues including those concerning the régimes of the high seas, the continental shelf, the territorial sea (including the question of its breadth and the question of international straits) and contiguous zone, fishing and conservation of the living resources of the high seas (including the question of the preferential rights of coastal States), the preservation of the marine environment (including, *inter alia*, the prevention of pollution) and scientific research'.

[2] It had already been agreed that there should be some increase in the size of the Sea-bed Committee. Because of the important decisions to be reached, there were strong pressures for representation from many countries. The resolution proposed an increase of 39 (to 86). It was subsequently raised again, first to 90, then to 91. The subcommittees were thus increased to the same size. Even by the standards of UN, these were pretty elephantine but their size reflected the importance attached to the subject.

reached to call a conference. This would decide both the question of sea-bed boundaries, as Malta had long demanded, and the provisions of the régime itself, as others wished. It had been recognized that the two questions, the limits and the régime, were inextricably intertwined, and that the two must be considered in conjunction. Finally, it had been accepted that the Conventions of 1958 might need some amendment, both as a result of technological and other developments, and because they had been formulated before nearly half of the existing members of the international community had acquired independence. And the preparatory work for convening such a conference, including the drafting of the necessary conventions, had been put in motion.

But the views on the individual questions that were to arise at the conference were as divided, and as mutually irreconcilable, as ever. Few could feel optimistic that, even when it did finally meet, the conference would be in a position to reconcile these deep and manifold disagreements.

The Preparations for the Conference

These underlying differences of view concerning the main purpose of the conference re-emerged as soon as the preparatory work began.

The first, and the most important, concerned the vexed question of 'priority' between the régime and the limits: should the limits be decided only *after* the régime, or vice versa? This must be reckoned one of the most profitless and puerile disputes in diplomatic history. For it should have been immediately apparent that neither viewpoint possessed the whole truth. It was equally valid to argue—as some rich countries did—that it would be as impossible to agree on the régime until one knew what area it was to apply to as that it was impossible to agree the limits without knowing what kind of system was to be established beyond them. The fact was that the two questions were inextricably interrelated. It was quite impossible to decide one definitely without knowing the other. The obvious and common-sense solution was therefore to seek to *discuss* the two in parallel, and to *decide* them together. This was roughly what the 1970 resolution calling for the conference had tried to achieve. Its text gave no explicit priority to one or the other.[1] It called for the new Committee to draft treaty articles embodying the régime and to draft articles on subjects and issues relating to the law of the sea: each of these could include references to limits, which were vital to both. To make the position doubly clear, the Canadian delegate, in introducing the resolution in the 1970 Assembly on behalf of the co-sponsors, had deliberately made a statement to clarify the meaning of the resolution on this point: both could be considered together, but a final decision on limits might have to await a decision on the régime. This did not prevent a number of delegations, especially Latin-American, from continuing to contest the position with great fervour within the Sea-bed Committee and so delaying its work for the next year and a half.

The other main dispute concerned whether it was necessary for the Committee to *complete* the preparation of the 'comprehensive list' of issues to be discussed by the conference before beginning to start work on drafting articles. To insist, like some Latin-American delegates, that the list must be completed first was of course yet another way of holding up substantive work on the various legal issues, including the definition of the area. Other developing countries were not prepared to go along with this obstructive view. Still less could the developed countries agree. Agreement was achieved finally on a programme, proposed by the chairman, which stated: 'it was understood that the Sub-Committee might decide to draft articles before completing the comprehensive lists of subjects and issues relating to the laws of the sea'. Even so, in practice the Committee did virtually no drafting till the list was completed eighteen months later.

The preparation of the list of subjects for the conference was in fact largely irrelevant: a choice of subjects would be made automatically in deciding which articles to draft. But the long discussion on the list perhaps served to ensure, before any practical work began, that all the various subjects to which any nation attached importance would be covered. It also allowed time for a general presentation of views and more thought and preparation in each government. The length of the lists proposed progressively expanded. The Soviet Union had suggested originally only two

[1] Its co-sponsors, however, recognized 'a certain priority' for the régime 'in the sense used by the International Law Commission': this meant simply that the final text on the régime might need to be agreed before the limits were adopted.

subjects. In July, 1971, Bulgaria put forward a formal proposal including five items. Norway proposed a list containing 15 items. The Latin Americans proposed a list with 9 main headings, but including altogether 48 items under these headings, some of them of a very broad character. Finally, the Afro–Asian group produced a list comtaining 54 items, grouped under 20 main headings: this included, however, one or two items, such as 'training, sharing of knowledge and transfer of technology', which could scarcely be considered questions of the law of the sea at all.

Eventually, in the summer of 1972, after much haggling, agreement was reached on a mammoth list of 25 broad items, with 90 subheadings, including virtually all that anyone had ever proposed.

Almost two years had been spent in these largely useless preliminaries. Only then could the real business of drafting begin. Draft Articles began to be put forward more frequently in 1973. The work of each subcommittee became more intensive. But all were far from agreement even on a single Article. It was hoped that these difficulties would be suddenly resolved at the conference itself. This, it was decided, would be held in Chile, in 1974, preceded by a procedural session in New York. As a result of the coup in Chile the location was subsequently changed to Venezuela where the first session of the conference began in June.

But it began to be clear that it would need perhaps not a single conference on the law of the sea but a whole series of them before agreement would be reached.

The Issues for the Conference

Let us now, then, examine the main questions on boundaries that are at stake at the proposed Conference and the widely conflicting attitudes of the principal powers towards them.

The fact that the conference was to consider such a broad range of questions, though it obviously complicated the task, had some advantages. For it was increasingly recognized that the entire complex of issues, diverse and difficult though they were, were closely interrelated. Every country, and every group of countries, had a different, sometimes conflicting range of interests within ocean space: either on the surface of the sea, on the bottom, or both. Agreement was likely to be reached only if all these varying interests could be accommodated through a vast, comprehensive package. The outcome was thus bound to be a compromise;

in which something was given, say in preferential fishing rights, to those countries which were mainly interested in fishing; something to the land-locked and shelf-locked countries without any waters or shelf of their own; something to the archipelago countries with special economic and security problems to protect; something to the highly developed, which already had the capacity to exploit, and must be accorded some opportunity for exercising it; and something to the developing as a whole, which reasonably felt that, if the resources at issue belonged to the whole world equally, the essential point was to ensure that all obtained a return from them. Only if these varying interests were balanced would a solution be possible. And only if all the manifold issues were considered together in a single mammoth negotiation, so that a concession on one point could be balanced by a concession on another, were the conflicting interests likely to be reconciled.

There was thus increasing recognition that the law of the sea as a whole would need to be re-examined. But in this case, should the four Geneva Conventions be abandoned? We have seen that the rich countries, and above all the Communist states, maintained for long that there was no need to reopen the questions they claimed had been already decided in 1958 in the Geneva Conventions. They contended that those Conventions were not simply the invention of the states which happened to be represented at that time, but embodied international customary law as it had developed over many years (the Soviet Union pointed out that in fact there had been more developing countries than developed at Geneva). The Latin-American and other poorer countries asserted, on the contrary, that the Geneva Conventions reflected only the opinions of those states which had been present and had subsequently ratified them (a number which varied between 25 and 47 states for the four Conventions), and had no wider significance. They could not thus in any circumstances be regarded as 'customary law'[1] binding on all existing members of the community.

[1] But Mexico, a maverick in this as in much else, held that the 1958 Conventions 'constituted an almost complete code of the law of the sea and had brought order into utilization and exploitation of an environment in which the interests of all nations clashed daily. They had given stable and precise form to many rules of customary international law which had been in existence for over three centuries. . . . The Committee's task was to supplement but not to rewrite them. . . . There was no justification whatsoever for treating those conventions as though they were an obsolete piece of 19th century leglislation'.

As so often, both arguments represented crude over-simplifications. Parts of the Geneva Conventions, especially those on the high seas and the territorial sea, could legitimately be regarded as having codified traditional customary law. They did not therefore depend on the number of countries ratifying. Others, and especially most of the Convention on the Continental Shelf, the one most directly relevant, could by no stretch of the imagination be regarded as doing so: that Convention had been the invention, and a much contested one, of the International Law Commission, which itself changed its mind several times and remained divided on the issue to the end.[1] Here, therefore, validity did depend on the number of ratifications. Moreover, to agree to look again at some of the subjects covered in the Conventions does not necessarily mean that they all had to be radically changed and re-written. But there had certainly been some revolutionary technological and other changes—greater threats of pollution, especially from oil; the growth of highly mechanized deep-water fishing fleets; increasing danger of overfishing and depletion of fish stocks; new weapons technology and new threats to security; above all, the increasing possibility of exploiting the sea bottom. The views of many governments had changed in consequence. The arguments in favour of some reconsideration now were therefore powerful. Thus, gradually, and despite the obvious complexity, the need for an over-all review of the law of the sea came to be more widely accepted, even by many of those who had been at first opposed.

This, however, raised a third major question: what was to be the future status of the Geneva Conventions as a result of the new conference? Suppose a two-thirds majority, mainly of very small developing countries, forced through at the conference a radically new law of the sea, perhaps extending the rights of coastal states in revolutionary ways, and restricting freedom of navigation in a manner unacceptable to the developed? Could countries which had ratified the Conventions, embodying what were the most widely accepted principles of international law at that time, and accepted by many as the existing state of the law, declare that they intended to stand by that commitment, claiming they were abiding by the rules of traditional international law as hitherto established, and as still supported by most of the major maritime powers of

[1] The proposed text was passed by a minority vote of the International Law Commission.

the world? In other words, could there, as a result of the forthcoming conference, emerge a situation in which there were two rival laws of the sea, each claiming equal validity—and which would come to be regarded, perhaps increasingly, as one law of the rich and one of the poor? This was by no means a remote possibility. It of course argued, like the similar danger in relation to the régime (see page 296 below), for the need to make every possible effort to arrive at consensus. But this did not mean that consensus would be achieved.

The first major question for the conference would clearly be: what should be the limit of the *territorial sea*? Here there had been a considerable change from the situation as it had existed in 1958. As the delegate from Peru happily declared: 'there was a clear and irreversible trend towards the extension of national jurisdiction. . . . The trend was not peculiar to Latin America, and the narrow limits . . . were not as sacrosanct as had been claimed in 1958'. Whereas in 1958 only thirteen countries had had a limit of twelve miles or more, in 1970 at least fifty-four already claimed twelve miles, and some claimed more still. Thus it was now virtually certain that there would be no agreement on a limit of less than twelve miles.

On the other hand, it was less likely than before that there would be a total breakdown on the issue. There was a considerable chance that there would be a large majority for a twelve-mile territorial sea. Among the 42 states which claimed less than twelve miles, a considerable proportion had announced that they would be willing to accept a twelve-mile maximum for the sake of agreement, and in practice almost all would do so. Moreover, the 30 or so land-locked countries, which had had little say in 1958 but would be represented at the forthcoming conference, had an interest in narrow boundaries to maximize the international zone: they would also probably support such a limit (since nothing narrower was feasible). Thus there was a considerable likelihood that there would be well over 100, and possibly close to 120, states favouring a twelve-mile limit at the conference. Against this, there were still only about 16 states which explicitly claimed more than twelve miles of territorial sea, though there were a dozen others which claimed wider jurisdiction for certain purposes.

It was of course precisely because of their awareness that they might find themselves isolated, that the wide-limit countries resorted to so many devices for postponing discussion of the matter. There was, however, a glaring illogicality in the position of the wide-limit countries. Their concern, and their wide limits, were, in almost every case, not related to the primary purpose of the territorial sea: full national jurisdiction of a general kind. They were interested almost entirely in fishing rights but believed that a claim to full jurisdiction could help them protect these. But fishing rights would certainly be considered in the conference as a quite separate issue from that of the territorial sea. So far as the sea-bed was concerned, their interests, as developing countries, required a narrow limit, to maximize the international area (this applied above all to the west-coast Latin-American states, which had little shelf anyway). Provided that their interests concerning fishing rights could be satisfied by other arrangements, therefore, they should logically have sought a *narrow* territorial sea. This could have been the basis for a deal. But they showed not the slightest readiness for a compromise of this kind. Far from seeking to separate the issues, they mainly continued to demand a single blanket limit of jurisdiction for all purposes.

There was another consideration which might have induced them to modify their position. Once the conference had taken place, their position under international law would be radically weakened. For twenty years they had been enabled to maintain their claim that every nation had the right to choose its own territorial sea because there existed no internationally agreed limit, and every government manifestly had made its own choice in the matter (a few countries, such as Britain, still maintained that the three-mile limit was a 'customary rule' of international law, but, as the proportion of states claiming wider limits steadily increased, this contention looked less and less convincing). But once the twelve-mile limit had received the endorsement of an overwhelming majority of states and had been embodied in an international Convention, it would become generally regarded as a rule of international law. And in these circumstances, such nations as the United States and the Soviet Union might feel that they would be justified in protecting the rights of their own fishermen to fish everywhere outside the generally agreed limit by every available means at their disposal: in other words, the U.S. Navy might be put into action. This was a sobering thought and was no doubt one of the reasons for the determined efforts of the Latin-American states to postpone any final international discussion of the question.

But the question of territorial waters was not only important in itself. It also brought in its wake a whole range of other difficult questions. One of these concerned the so-called *right of 'innocent passage'* within such waters. This was a traditional right going back to time immemorial. Article 15 of the Convention on the Territorial Sea laid down that 'the coastal state must not hamper innocent passage through the territorial sea'. On the face of it this meant that foreign ships had an unfettered right to sail there. But in practice it was not always unfettered. The coastal state could suspend temporarily the right of innocent passage in specified areas for 'security' reasons, provided that this was not a strait 'used for international navigation between one part of the high seas, and another part of the high seas or the territorial seas of a foreign state'. In these international straits the rights of innocent passage were not to be suspended even for security reasons.

Most developed countries held that this right of innocent passage extended not only to merchant ships but to naval vessels as well. Under the Convention on the Territorial Sea, a warship could only be required to leave the territorial sea if it did not 'comply with the regulations of the coastal state concerning passage through the territorial sea'. International law itself was somewhat uncertain on the point. According to one well-known authority: 'a right for the men of war of foreign states to pass unhindered through the maritime belt is not generally recognized. Although many writers assert the existence of such a right many others emphatically deny it' (a situation in international law, alas, not uncommon). But the same authority adds that 'a usage has grown up by which passage which is in every way inoffensive and without danger shall not be denied in time of peace' and that 'it is now a customary rule of international law that the right of passage through such parts of the maritime belt as form parts of the highways for international traffic cannot be denied to foreign men-of-war'.[1] Not all accept this opinion even for those areas. Within the Sea-bed Committee, a number of developing countries challenged the view that warships could share the right of innocent passage, whether in international straits or elsewhere. Even Canada said that it felt that the traditional concept of innocent passage needed clarification and redefinition. But the big maritime countries,

[1] Oppenheim-Lauterpacht, *International Law*, 8th ed. 1955, Vol. I, p. 494.

especially the United States and the Soviet Union, were determined that the right of passage by warships, at least in all international straits, should continue to be protected.

This then raised another controversial point: the right of *passage through straits*. This arose automatically from any new agreement on the territorial sea. A twelve-mile limit, if agreed, could bring the whole of a large number of international straits (any less than twenty-four miles wide) within national waters. The Soviet Union claimed that there were 100 straits which would be 'closed' in this way. It was therefore suggested that special measures would need to be taken to preserve free passage in such straits. There was not, however, any agreement on how this freedom of passage should be preserved, if at all. In theory it would be partially protected by the right of innocent passage which existed within all territorial waters. But the big maritime states, especially the United States and the Soviet Union, felt that in the case of international straits this needed special reinforcement, especially because, as we have seen, even innocent passage was not unchallengeable. The United States therefore introduced in the Committee a draft article, which provided for the right of 'free transit' through such straits. This allowed the coastal state to name a corridor which could be always available to international shipping or for international airways if necessary. These provisions were not much liked in the Committee: a few states controlling straits explicitly denied the need for a special régime. Other countries expressed their concern to maintain sovereign discretion on the question. Here was another important question on which there remained ample room for controversy.

Here too, as in territorial waters generally, the question arose whether any special rights would apply to naval vessels: in the Corfu Channel case, the International Court had held that a channel of this kind must be held open for innocent passage without previous authorization, even by naval vessels. Before the 1958 Geneva Conference, the International Law Commission proposed that the passage of warships through territorial waters should be subject to prior notification or authorization. A proposal requiring authorization was passed in committee in that conference, but failed to obtain a two-thirds majority. The delegate of Malaysia, which controls an especially important international strait, declared in the Sea-bed Committee that his government would demand prior authorization for any such passage of warships; and subsequently Malaysia and Indonesia jointly declared that they would demand prior authorization of all passage of warships in the Straits of Malacca.[1] Here too, there was ample room for debate, with the big naval powers against most of the rest.

Another question related to that of the territorial sea concerned the position of *archipelago countries*. Both Indonesia and the Philippines have insisted strenuously on the right of such countries to measure their territorial sea from the outermost points of their outermost islands, so enclosing huge areas of ocean within their internal or territorial waters. Indonesia first declared this doctrine in 1957. The Philippines' delegate declared that 'the Philippine government considers that the waters between those islands form part of the archipelago and that it was vital, not only to its economic life but also to its security, that it should exercise control over them. It further laid claim to the sea which, historically, had always belonged to the Philippines, just as other states laid claims to bays and gulfs for historical reasons'.[2] Limits already claimed were 'a reality which nobody could ignore and the existing realities should serve as the foundations for the rules of international law' (in other words, the more outrageous the claims made by a state in the past, the more it should be allowed to keep now). The Indonesian delegate, defending the same system of baselines, said that the Indonesian 'islands, of which there were more than 13,000, and the intervening waters, formed a single unit. For the Indonesian people and government that was an axiomatic fact of life. It was also an economic necessity, for it was from the sea that from time immemorial the inhabitants of the Indonesian islands had drawn subsistence.'[3]

What rules can such a state reasonably apply? The question at issue here is not the breadth of the waters (both Indonesia and the Philippines accept the twelve-mile limit) but the way in which the

[1] In March, 1972, Indonesia and Malaysia announced that the Straits of Malacca, which in places are only six miles wide, were being taken under their control and that vessels would enjoy there only the right of innocent passage. Passage of naval vessels would require 48-hours notification (this has apparently been ignored by the United States and other governments) and the passage of tankers over 200,000 tons would be prohibited to protect navigational and pollution dangers. Western States have held that the waters are international, and Japan, which is willing to pay the cost of dredging and other improvements, would like them internationally administered.

[2] A/AC138/SR55, p. 128.

[3] A/AC138/SR55, p. 132.

baselines are drawn. This can affect both continental shelves and territorial seas. The 1958 Convention on the Territorial Sea had laid down that a baseline of twenty-four miles can be drawn across bays in such a way as to enclose the maximum area of water, except in the special cases of 'historic bays', and of the mouths of rivers, where they may be longer. Otherwise, baselines must not depart from the general direction of the coast, and the sea areas must be sufficiently linked to the land to be subject to the régime of internal waters.[1] How such a principle should be applied to archipelago countries, or what other rule should be applied there, had never been decided by the International Court or in international law. The longest baseline used by Indonesia is 190 miles, and it was unlikely that the principles declared by Indonesia and the Philippines would be generally accepted.

In any case, any principle relating to archipelagos and island groups would require a definition of an *island* for this purpose. Such a definition is also required, and urgently, in relation to the continental shelf: otherwise, as can be seen, nations might claim enormous areas of shelf which were dependent on tiny specks of uninhabited rock. The 1958 Convention on the Territorial Sea had itself given a definition of an island for the purpose of that Convention: 'an island is a naturally formed area of land, surrounded by water, which is above water at high tide'. But it does not necessarily follow that the same principle should be used in defining islands for the purpose of the continental shelf (even for the territorial sea the definition might be changed at the forthcoming conference). The importance of this issue was highlighted by Dr Pardo within the Sea-bed Committee: 'Virtually uninhabited islands could give their respective possessors the right to claim jurisdiction over millions of square miles of ocean space. The possession of St Paul's rocks, Fernando Noronha, Trinidade, and Martin Vaz, all virtually uninhabited, could more than double the potential claims of Brazil to ocean space. His delegation had calculated that potential claims based on the possession of islands inhabited by a total of less than 3,000,000 people could cover nearly two-thirds of world ocean space'. One criterion might be to include only inhabited islands. It might also be necessary to distinguish between islands which are independent states or will shortly become independent states and those which will remain

[1] On this basis, the International Court, in the Anglo-Norwegian Fisheries case, accepted baselines as long as 45 miles.

dependent states.[1] Finally there is the question of whether isolated islands very far from a mainland (such as Rockall, claimed by Britain for this purpose) can be classed as islands for the purpose of fixing limits. The recent violent disputes between Japan, Taiwan and China over rights to the Tiao-yu (Senkaku) Islands, in the China Sea, which are totally uninhabited, and little more than sand-banks, is an indication of the importance which such matters can have, once it is believed that they control continental shelves which may be rich in resources. This is therefore another issue which could be of vital importance, both to individual nations and to the viability of the international régime, and will thus be bitterly debated at the conference.

Next, there was the question of *closed or semi-closed seas.* Certain states, especially Mediterranean countries, insisted strongly on the need for some special arrangements for the sea-bed within such seas. In some cases this concern related to fishing rights on the surface as much as to resources beneath the sea. As the representative of Iran pointed out, 'the marginal seas were geologically part of the continental land mass: biologically they belonged to the same eco-system as the neighbouring land mass; economically they came within the fields of socio-economic gravity of the riparian communities, which were increasingly dependent on resources of the sea for their subsistence and economic development. The intrusion into those seas of fishing fleets of distant states created an anomalous situation which caused serious economic dislocation. . . .' He therefore hoped that the Law of the Sea Conference would examine the 'shortcomings of international law' in that respect. And this became ultimately one of the subjects on the agenda for that Conference.

Seas or lakes that are totally surrounded by the territory of a single nation, international law has agreed, become part of the territory of that nation and under its full jurisdiction. Similarly, where seas or lakes are totally surrounded by the territory of several states (like the Caspian Sea, Lake Constance, or Lake Geneva) they may through agreements among such states be divided up among them (though there are some international

[1] The 1971 legislation in Britain to make Rockall an integral part of the United Kingdom was presumably designed to prevent it being classified as a colony or as an isolated island whose shelf was not Britain's. One of the questions on the agenda of the great Conference is the 'régime of islands', including 'islands under colonial dependence or foreign domination or control'. There is also the question of islands, such as the Dodecanese, in the shelf of another state.

lawyers who in this case claim that there are 'high seas' in the middle, even if they cannot be reached from the oceans). But the 'semi-closed' seas such as the Baltic and the Mediterranean are not really in any way fundamentally different from any other sea. Nor have they been treated differently under international law until now. For a time during and after the Napoleonic wars, the countries of the Baltic tried to maintain it as a 'closed sea'. But this was resisted by Britain and other states; and the Treaty of Versailles definitely established 'free passage into the Baltic to all nations'. Similarly, Russia for a time sought to regard the White Sea as closed, but this claim too was resisted by other nations. The Italian delegate in the Sea-bed Committee attempted to show that, geologically, the shelf of a sea such as the Mediterranean was different from that elsewhere and could properly be regarded as in a special category. This may, however, be true of many other parts of the ocean bed. And there seems no sound reason why the principle finally adopted for defining the continental shelf, whether based on depth or distance, should not be applied equally in these areas as in the rest of the oceans.

Another important question concerns totally new types of jurisdiction. Canada and Iceland, for example, have strongly affirmed the right of states to take action in the high seas to prevent *pollution*. And Canada has unilaterally asserted powers for this purpose. In the words of the Canadian delegate, 'because the coastal states suffered the most drastic effects of marine pollution damage, future international law would have to recognize their fundamental right to protect themselves against that threat to their environment'.

Here, as in the case of fishing and conservation, the real question is not whether action against pollution is desirable or necessary, but whether it should be taken *unilaterally*. If there is an inter-relationship between pollution in coastal waters and in the high seas, does this imply a right for the coastal state to undertake action even in the high seas? Or, conversely, for international regulation to extend into the coastal waters? It is reasonable to assume that as a result of the Stockholm and the 1973 IMCO conferences, and perhaps of new conventions drafted by the present Conference, there will shortly be more effective international regulation of marine pollution. The type of measures that may be taken, by both coastal states and others, may be clearly laid down. The problem that then arises is how far a coastal state would be entitled to interfere with traditional rights of free navigation in the high seas to enforce *higher* standards than those internationally applied (especially when such regulations can very easily be used as the justification for action taken on quite other grounds). In other words, should the international standards be not merely a minimum but a maximum too? Alternatively, if the international standard is laid down in a Convention, should the coastal states be given the right to enforce it on behalf of the international community? Finally, what happens about shipping countries which fail to ratify an international convention, and so choose to remain outside the international system? What right has the coastal state in relation to such ships? A considerable conflict of views took place, within the Sea-bed Committee and outside, between those, such as Canada and Iceland, which maintained the right of unilateral action, and the traditional maritime states, such as Britain, the United States and Russia, which resisted such a right. There is no doubt that there is at present wide public sympathy for any measure that is claimed to be directed against pollution. And it is certainly the case that if regulation is inadequate, it is the coastal state that suffers the worst damage, as Britain itself has good reason to know. For this reason, coastal states have already been given, partly on Britain's initiative, the right to take action against shipping outside their waters which is in danger of polluting the coast in the case of major spillages. Coastal states may claim more sweeping rights to institute measures against pollution within a certain distance from their coasts, perhaps in some new 'contiguous zone' created for the purpose. Again there would be much dispute about this.

This leads to another major type of jurisdiction for the coastal state, which has been a source of major conflict in recent years: over *fishing rights*. If the territorial sea was extended to twelve miles, exclusive fishing rights would be extended automatically for the same distance (though this could be qualified by special agreements on a regional basis, as has happened in West Europe, giving some other states rights even within those waters). In fact already many three-mile countries claim fishing limits of twelve miles. The question is really: can the coastal state claim rights, preferential or exclusive, beyond that point?

Many poor countries would certainly have said, 'Yes'. But the developed countries were generally critical at first of claims to wide fishing rights. This applied above all to the Communist states. The Hungarian delegate declared that 'a great deal had been said about the rights and interests of coastal states but some of those states had been extending their national jurisdiction to a degree that could be described as excessive'. And the Soviet representative stated that 'if claims to exclusive rights over extensive off-shore regions of the high seas were admitted, a huge quantity of food resources would be lost. Research had shown that in the South West Atlantic, nearly 12 million tons of fish could be caught every year without affecting the process of reproduction; but the coastal states which had established a 200-mile limit for their territorial waters were catching only 250,000 tons a year, or barely 2 per cent of the total potential catch in that area'. However, these attitudes were gradually modified. Many developed countries have now accepted wider rights for coastal states. The Soviet Union and Japan, representing the interests of distant-water fishing countries, proposed a preferential right for the coast state but limited to the proportion of the total stock its own fleet could catch. Canada proposed that the coastal state should be accorded certain rights as the 'custodian', under internationally agreed rules, for the world community, including a preference which might involve exclusive rights for some species. The United States, Australia and New Zealand proposed arrangements, also favourable to the coastal state, giving it (again subject to its fishing capacity) virtually complete control over all except the highly migratory species (such as the tuna), which were to be governed by fisheries commissions.

These proposals still did not go far enough for some developing countries. There was increasing recognition of the importance of the fisheries issue. For some states it was this question that was most immediate and determined their attitude on all others. The Latin-American states in particular presented the issue as defence of natural resources which 'naturally' belonged to them against the depredations of imperialist invaders. In the words of Peru, the developing countries 'are successfully defending their rights to use the resources of their own territories and the contiguous zone for the well-being of their own peoples' while rich countries, such as the United States, resort 'to every kind of pretence to prevent the inevitable'. Many other developing countries agreed on the need for fairly broad fishing and other economic rights for the coastal states, but accepted the distinction, which the Latin Americans ignored, between territorial sea and fishing zone.

Such countries thought increasingly in terms of an economic zone, in which a coastal state would be able to enjoy special rights in relation to all economic resources, whether on the sea-bed or in the waters, but would not claim jurisdiction for other purposes. Kenya introduced a draft article formally proposing an 'economic zone' of a maximum width of 200 miles, in which the coastal state would exercise sovereign rights 'for the purpose of exploration and exploitation' of living and non-living resources, either on an exclusive or a preferential basis. Special arrangements might be made within each region to provide facilities for neighbouring land-locked states. Two Conferences organized in 1972 by Caribbean and African states endorsed this concept of an economic zone or 'patrimonial sea'. Many felt that this could be combined with recognition of a relatively narrow twelve-mile territorial sea. Even Iceland said that 'it was not necessary to insist on a wide territorial sea if fisheries jurisdiction was adequately dealt with'.

Some were in favour of a regional system of fishing preferences. Some Latin Americans had always favoured this: above all, Uruguay (which might so have gained access to the rich grounds of its neighbours). Such a system could in fact be just as protective and discriminatory as control by national governments. Moreover, it is clearly better that there should be uniform rules all over the world. Here the same question arises as over pollution: if conservation measures were needed, should they be taken unilaterally, whether by a single nation or a region; or only by some international body set up for the purpose? Should they be objectively examined and authorized by the fisheries commission responsible for each stock of fish, or by the Fisheries Committee of the FAO, rather than by the interested parties themselves? Should the fishery commissions be under an obligation to grant preferences clearly laid down, or even exclusive rights, to the coastal state or region in particular areas or circumstances?

Finally, even if the idea of a preferential or exclusive zone was agreed, how far should such a zone extend? Should it be measured in miles from the shore, or in terms of particular stocks of fish? What degree of preference should be accorded to the coastal state? Should there be an exclusive *and* a preferential zone? Should the zone be related to the physical continental shelf? This is what Iceland asserted in extending fishing rights to the edge of its own continental shelf in 1971. Or should it be related to a theoretical shelf, measured in distance? This is what the

Latin Americans maintained in asserting a limit of 200 miles for fishing and sea-bed resources together, and what other nations in Africa and elsewhere supported in calling for a 200-mile 'economic zone'.

This led to the most fundamental question concerning limits, the one that we have come across so often in this book: what should be the outer limit of national authority over the resources of the sea-bed? For this would have the most effect in determining who would get the benefits of sea-bed exploitation. The question was still often spoken of in terms of the limits of the 'continental shelf'. But there was much to be said for abandoning altogether the use of that traditional but quite misleading phrase. Because of the fickleness with which nature has endowed different countries with narrow or broad shelves, or with none at all, a limit based on the geological concept—that is, depth—was arbitrary, ill-defined and irrational. It became increasingly logical to think rather in terms of a zone of jurisdiction—either a general economic zone, or a sea-bed resources zone quite unrelated to geological formation. This could be measured in the same way as other such zones: in distance from the coast. And during the course of the discussions, increasing support was expressed for using a measure of distance to determine the line, either in combination with that of depth, or as the only measurement.

Whatever basis was used, the fundamental question remained: how narrow, or how wide, should the limit be? Here arose the most fundamental and important conflict of the whole discussion. For it determined how sea-bed wealth would be divided. The nations with the greatest interest in narrow boundaries were land-locked and shelf-locked countries, and those with no shelf. In this way they would maximize the revenues for the international régime through which alone they could benefit. Nepal, Afghanistan, Singapore, and others expressed themselves eloquently in this sense. The Afghanistan delegate logically proclaimed his country's 'right of inheritance to what had been internationally recognized as the common heritage of mankind'. He declared that 'any unilateral action by coastal countries regarding the exercise of their national jurisdiction over the sea was incompatible with the concept of the common heritage of mankind'. Other land-locked states expressed themselves similarly. Together they proposed a limit of forty miles for the national area, with some powers of control beyond that line for the coastal state. The coastal countries of course had an opposite interest—in a wide boundary, to allow them to appropriate the largest possible proportion of sea-bed resources for themselves. 'In the exercise of their sovereignty', as the Peruvian delegate put it, 'the coastal states had to establish the limits of their national jurisdiction . . . taking account of . . . the need to benefit from their resources'. He did not define in what sense, and to what extent, sea-bed resources were 'theirs'.

A few internationally minded states favoured a relatively narrow limit. Sweden declared that 'Sweden strongly favoured setting aside for the international community an area as vast as possible, so that the developing countries could benefit from the profits of the envisaged exploitation of the resources of the sea-bed'. Most of the other rich countries appeared to favour wide limits: Britain in particular made no secret of its preference in this respect. The United States was in an ambiguous position: it favoured narrow limits for certain purposes—for example, pollution control and restriction of navigation. But for the most important purpose, control over exploitation, it proposed originally wider limits than any other country: the edge of the continental margin (about a fifth of the entire sea-bed and the part richest in resources). The Communist states, apart from their repeated support for the Geneva Convention, did not declare themselves in favour of any specific distance or depth: having few favourable coasts, they were perhaps concerned rather with obtaining the greatest possible freedom of activity in the international area than with wide limits as such.

Everybody agreed that the decision on the limit would be influenced by the nature of the régime established. And this in turn induced some slight moves towards compromise. The U.S. delegate, for example, declared that 'if the trusteeship adequately accommodates coastal and maritime and international interests, the method of delimiting its outer boundary should not be an obstacle to agreement'; against this, the delegate of Ghana stated that 'if agreement could be reached on a strong régime and machinery, a fairly large number of states would accept relatively narrow limits of national jurisdiction'. Even more likely was a deal in which the rich countries accepted a fairly broad economic zone beyond the territorial sea in return for a territorial sea of twelve miles only. Certainly the increasing recognition that all issues were interrelated suggested the need for an over-all compromise of this sort. But on this particular question, the most basic of all, such a compromise would not be easy to attain.[1]

These then were the principal issues discussed during the preparations for the great Law of the Sea Conference. There was one point at least which began to emerge more clearly during these discussions. Increasingly, the emphasis shifted away from the traditional concept of two rigid dividing lines, one at the top and the other at the bottom of the sea; between one area that was fully national and another that was fully international. In two different ways this line became increasingly blurred. First, there might in future not be one all-important boundary line or even two, but a succession of limits of jurisdiction for different purposes: for general jurisdiction, for customs and sanitary control, for security, pollution regulation, fishing control, and sea-bed resources, for example. More important, in each of these spheres there would be no rigid line between one side that was national and the other that was international. States would have some international responsibilities—for example, concerning the control of pollution, the supervision of exploitation, the conservation of fish stocks, preservation of freedom of navigation, research, and others—even in the areas over which they retained primary control, and would be subject to international rules over them. Conversely, the international area beyond would no longer be an area of total and unlimited freedom: it would be one in which both national states and international bodies would each accept certain clearly defined responsibilities—for example, on pollution, rights of exploitation, conservation of fishing and other matters.

It was the exact balance between these various rights and responsibilities, between national control and international regulation, which would need to be decided in the forthcoming discussions. Given the complex and conflicting interests of the many states and groups of states involved, it would be no easy task to secure agreement on the very many difficult questions at issue.

[1] For a suggestion of possible overall compromises, see pp. 264–67 below.

CHAPTER 10

The Régime

Types of Régime

While these debates on the principles and the limits were proceeding, another was taking place on a still wider question, the régime itself. What kind of system should be applied within the international sea-bed area? Let us look at the broad types of 'régime' that have been discussed and the main characteristics of each.

What was meant by the word 'régime' in the first place? The word was an extremely vague one. It meant different things to different people. To some it meant simply a set of rules—for example, the obligation to prevent interference with navigation and other uses of the seas, to avoid pollution, to observe safety practices, and so on. So long as these basic principles were observed, the sea-bed and its resources could be used by all who were able to do so as they wished. This seems to be the kind of system which the Russians, for example, and other minimalists, understood by the term 'régime'. And some of the developed western countries at first seemed to favour something only a little more elaborate.

To the majority of countries, however, the word implied a system very much stronger than this. It denoted, at the very least, an effective international authority to administer the area; principles for deciding who could exploit what, and when and where; a range of obligations to be observed by states in undertaking exploitation; and a system for distributing revenues derived from exploitation. And already, after the first year or so of discussions, because of the majority pressures, and perhaps of greater understanding of the subject, almost all countries began to accept that at least something more than a mere set of rules was required.

Even so, a wide range of 'régimes' was possible. The simplest of all would have been a system of *notification* alone. If any government, or company, discovered resources on or under the sea-bed, and wished to exploit them, they would be obliged to inform a central authority of their intention. This could act as a confirmation of the claim, and would prohibit others from undertaking exploitation within that area. It would thus preserve a minimum of order in the oceans. It would have the effect of reducing the uncertainties which would otherwise develop in an area subject to any claim from anywhere, and give confidence for investment. Additional features could be added if desired. Rules regarding pollution, safety, freedom of navigation and so on could be laid down; though these might be left to national governments to enforce on their own companies. There could even be some provision for payments to be made to an international body, or for international purposes, perhaps on a voluntary basis.

This system came about the lowest within the spectrum of theoretically possible régimes. Most states, especially the developing countries, held that it would be quite inadequate. It would represent only a cosmeticized version of 'first come, first served'. There would be no effective international control over resource management, pollution and so on. But its outstanding disadvantage was that, even if it included provision for the payment of royalties for international purposes, the actual use and exploitation of sea-bed resources would in practice be confined to a very small number of highly developed countries (possibly only one), which alone possessed the necessary technology to undertake it. If such countries had only to notify but not obtain permission for exploitation, they would not be effectively controlled; and they would increasingly take over the whole sea-bed for their own use. The preponderant role and the innumerable advantages which highly developed states anyway held within world economic activity, through their higher technology and larger investment resources, would be still further increased. And the main return from assets, which were essentially international in nature and the common property of all, would be appropriated by a few advanced countries and companies which had the least need for additional revenues.

A slightly stronger international system would involve not merely the notification of claims but their *registration* by a central authority. Again, a range of possibilities existed. Registration might be virtually automatic once a claim had been notified; in which case the system would be little different from notification, with all the same disadvantages. Or it could depend on, say, evidence of a genuine find, a commitment to work at a certain rate of production, or evidence of adequate technology and financial backing. But here, too, if the authority had otherwise no discretion to withhold registration, in effect the situation would still be little different. Any such system would provide a strong incentive for a race to register the maximum possible claims as early as possible, and lead to a kind of space-age gold rush all over the sea-bed to find and reserve deposits. Registration could, however, be selective; that is, even if all these conditions were fulfilled, the authority might refuse to register, on the grounds that a particular company, or nation, had already been awarded its fair share of the sea-bed resources. Under such conditions a system of registration, or licensing (for here registration and licensing would be synonymous) would amount in effect to a form of rationing. Only this would procure an effective system of international control. And this, increasingly, was demanded by many countries.

But such a system would still provide considerable advantages for some nations over others. For the *initiative* would still remain with the operators. These would seek out the most favourable sites all over the oceans before applying for registration. Only the most technically advanced countries and firms would be equipped to explore the ocean bed and so to find the best sites. This would still give them a huge advantage over those states or companies with a less advanced technology. The power of the registering authority would be restricted to the essentially *negative* one of preventing an operator from undertaking exploitation in a particular site, rather than the positive one of allocating sites chosen by the authority itself. This is a crucial distinction; and thus the important division among systems is not, as is often suggested, between registration and licensing; it is that between licensing of a negative or of a positive kind.

The next type of régime that could be adopted, therefore, would involve a system of *allocation*. Here the initial choice of sites would no longer lie with the operators but with the licensing authority, which would decide to make particular areas available for exploitation at its own discretion. A number of methods for allocating sites within a particular region would be possible. Sites or blocks could be auctioned, and made available to the highest bidder. This is roughly what the United States does in relation to its 'outer' or federal, continental shelf, and what Britain has recently done in the latest phase of North Sea allocation. In such systems some exploration of the area—for example, through seismic surveys—is usually allowed before the sites are auctioned. But in this case there would still remain substantial advantages to the most developed countries or companies: on the one hand, they have larger funds to bid at the auctions; and on the other, they are better equipped to explore sites effectively beforehand, and to know which they should seek to acquire. Even if *exploration* rights only were auctioned, this would be on the understanding that a company acquiring such rights would also have first option in acquiring rights of exploitation: even here though to a lesser extent, the country and company which is most advanced is likely to have the greatest opportunities of choosing, and taking advantage of, its exploration rights.

But allocation could also be based on a deliberate effort by the authority to spread blocks to different operators, either by limiting the number of blocks to each bidder, or by deliberately inviting bids for particular blocks from particular operators. This is sometimes done by Middle East states trying to spread their risks, or to reduce political dependence. In other cases, there is not only an attempt to spread the number of operators, but to invite particular operators to take up particular blocks; something like this was done by Britain in the Irish Sea, when British nationalized corporations were given special opportunities for participating in the exploitation of a proportion of the area. Finally, there can be a completely random distribution, or drawing of lots, for the blocks, so ensuring that all potential operators (or nations) have an equal chance. Though some may acquire profitable sites and some barren ones, this is left to fate to determine, and not to the existing level of technological expertise. This again is a system which has been adopted by some countries in leasing submarine rights. And it was this which to many states seemed to represent the fairest way of sharing out rights in the sea-bed.

Under any of these schemes the blocks would be leased only for a given period. Renewal would be subject to confirmation that it had been fully worked; that good operating practices had been observed; and that adequate safeguards against pollution, accidents and other hazards were being maintained. If desired, sites, or a proportion of them, could be automatically relinquished after a particular period, or if they were not being adequately worked. But even under this system there would be a need for a stable and reliable régime which would provide adequate incentives to operators, whether companies or governments. Unless there was some assurance that sites could be retained for a fairly substantial period, no company might feel it worth bidding. When a block was renewed, it might be under new terms: for example, the royalty demanded for the international authority could be progressively increased as the production potential became higher. So the authority could gradually increase its revenues.

The fourth possible type of régime was one under which all exploitation would be undertaken by an *international enterprise* on behalf of the international community as a whole. This would have the advantage that the benefit from sea-bed wealth belonging equally to all would be shared equally among all. The sea-bed would not need to be divided up among little national blocks spread all over the ocean. No special advantage would be acquired by the most technically advanced countries. And, for the first time, an international body would become engaged in a purely economic activity on behalf of the whole world community. It would have the disadvantage that it might require a large and cumbrous bureaucratic organization, without commercial experience or motivation, and without the incentive of the market to promote efficiency. It would prevent the exploitation of sea-bed resources by other enterprises, private or national, and so could reduce the benefits that the world economy might gain from them. And even the interests of developing countries might be damaged by the reduction of production, and so of international revenues. At present, most of the world's limited expertise in undersea technology is concentrated in a few large companies, mainly North American; and it is by no means certain that this expertise would be effectively mobilized under a fully international system of exploitation of this type.

On the other hand, it would be perfectly possible to have a system of international *ownership* of some or all of the sea-bed blocks, whoever was engaged to undertake the actual exploitation. In this case, companies could be engaged directly by the authority to perform the work of exploitation on a contract basis. Alternatively, an international exploiting agency, or regional bodies, might be formed to exploit some of the blocks in competition with private companies or government enterprises working others. Finally, it would be possible for an international resource authority to enter into joint ventures with other bodies, whether national or private.

This, in a much simplified form, was the main range of alternative régimes which came under discussion from 1967 onwards. But there was one other set of alternatives concerning the régime that must be mentioned briefly here. Whatever type of registration or licensing system was adopted, the blocks could be leased by the international authority in two ways: either to national governments (or to groups of governments or regions), which might then in turn lease to companies, taking a royalty themselves on the way; or direct to companies by the international authority itself.

The former system would have the advantage that every nation would acquire a *direct* benefit (through their own royalties), as well as an indirect benefit through the royalty for the international authority, from the exploitation of the sea-bed. It would give small and developing countries a sense of genuinely sharing in the control of the area through control of their own blocks. And it would enable governments to take some of the responsibility for supervising the operating companies and ensuring that they complied with the necessary requirements on pollution and other questions. In this way it would, it was said, reduce the scale of bureaucratic control. But it would have the disadvantage that it would create a further, third tier of authority, between the controlling agency and the operator. This would have no very obvious function in relation to the sea-bed. And it would spread the nation-state system right across the sea-bed, so creating innumerable little square colonies all over the ocean floor, operated by individual nations for their own national benefit. It would thus be a far less international system than one in which the international authority leased direct to the companies. Here too, therefore, an important range of possibilities was open in discussion of the future sea-bed régime.[1]

There was also, of course, room for wide disagreement about the type of institutions which would be required to maintain and enforce the régime. Virtually everybody—even the Russians—

[1] For further discussion, see pp. 224–29 below.

agreed that there should be some rules concerning safety measures, pollution, conservation and other questions. But there was far less agreement concerning how these should be enforced. Some held that this should be left entirely to national governments, either in relation to their own companies or to their own areas of the sea-bed. But others, including some developed western countries, accepted from the start that there would be a need for some international sea-bed agency to undertake this work. There would have to be some body to undertake inspection, to ensure that operators conformed with the standards which had been laid down. Such a body might also have a role in providing technical assistance to developing countries in sea-bed technology. But the exact powers of such an authority, and the voting system to be used within it, would clearly be a highly contentious question. Finally, there would probably need to be some kind of tribunal for settling disputes, both between nations and between nations and the international authority; and here too there was room for wide differences of views on the powers it should have.

These were, in very general terms, the kinds of issue that began to be discussed among the nations of the world when they sought to define the 'régime' which should be established within the vast expanses of the sea-bed now becoming open for use for the first time. The decisions reached could affect the disposal of a large part of the world's remaining wealth. Inevitably, different nations and groups of nations quickly lined up for and against one or other of the various systems. Equally inevitably, these positions closely reflected their own national interests.

The developed states on the whole tended to prefer the weaker systems, providing greater freedom (and profits) for exploiting companies; the developing preferred a more international type of régime, providing for countries of less advanced technology both a greater degree of control over exploitation and a greater share in the revenues generated. The Communists favoured the weakest possible system, with the minimum degree of international control. At the other extreme, some of the Latin-American countries, as we shall see, came out in favour of direct exploitation by an international enterprise, though at the same time wanting wide powers for coastal states.

But a description of group positions can be misleading. Almost every individual nation had a special interest of its own, and therefore a special position. No block, not even the Communist, was cohesive. To understand these varying view-points, it will be necessary to look in greater detail at the debate which has taken place on the question.

The Debate on the Régime

The discussion of the régime at first took place in parallel with the debate on the principles and the limits already described.

The first main issue concerned the nature and powers of the proposed sea-bed authority. The developed countries in general laid heavy emphasis on the undesirability of any large-scale or cumbrous machinery. They stressed insistently the need for any régime to provide effective incentives for investors; to possess 'effectiveness, credibility, and impartiality'. They felt it was important not to be precipitate or over-ambitious in seeking to establish any new international authority. In the words of the U.S. delegate, 'the establishment of international machinery endowed with excessive powers of control would, in this delegation's view, run counter to the objectives which all countries shared'.[1] These countries tended in general, therefore, to favour a system towards the lower end of the spectrum described earlier. The Communist countries insisted, even more emphatically than the western countries, on the need for caution. They accepted that legal principles must be worked out to govern exploitation and to 'foster the development of international co-operation' (to use their favourite phrase), to provide 'free access to the resources of the sea-bed to all countries' (as the Bulgarian delegate put it) and to ensure 'fullest freedom of the high seas'. Indeed, free enterprise was in many ways their watch-word. The Bulgarian delegate even demanded, in words which would have done credit to an arch-capitalist monopolist, that 'the régime to be established should be efficient and profitable with minimum operating costs'. Their concept of a régime seemed to be of a system in which national governments operated almost entirely independently, apart from general undertakings to observe certain rules, and to seek 'co-operation'. As to the nature of the authority, they deplored the tendency of the UN 'to have recourse to hastily elaborated institutional arrangements that had proved to be ineffective and inadequate'. Above all, the Communist delegates deplored any

[1] Economic and Technical Subcommittee, Summary Records A/AC138/SC2/SR21, p. 83.

idea of a supranational régime as unrealistic and impracticable, since it 'disregarded the existence of states with different social systems and types of ownership', and could even lead, so they somewhat implausibly declared, to the control falling into the hands of 'large-scale, capitalist, imperialist monopolies' (in fact the whole point of a régime was to control such monopolies). They thus inevitably favoured a system at the very bottom of the spectrum.

Representatives of the developing countries held the opposite view. They pointed out that inadequate international control would mean that only the most advanced countries could participate in the exploitation of the sea-bed. They regarded it as essential that there should be a strong central authority which carefully scrutinized applications to operate within the area, and demanded an adequate royalty for international purposes. They insisted, therefore, in the words of the representative of Madagascar, that 'future machinery should have extensive powers to enable it to perform its functions, which should include organization, administration, management and co-ordination of operations, . . . supervision and inspection'.[1] These countries therefore, mainly favoured a system towards the upper end of the spectrum of possibilities.

Again there were considerable differences on the method of proceeding. The developed countries, both of east and west, insisted that it was necessary to proceed cautiously and without too much hurry. They wanted an authority, with limited powers at first, which could be slowly built up and expanded as sea-bed production increased, and as the task became better understood. As the Japanese delegate put it, 'the establishment of international machinery was a complex problem and should be approached with caution'. The poorer countries were suspicious of this approach. They felt that an effective authority with wide powers should be set up from the start: otherwise it might be found, as the representative of Ceylon warned, that the provisional lasted only too long, and could not be easily changed. They were apprehensive concerning the speed at which exploitation of sea-bed areas was developing, and feared that they might find themselves faced rapidly with a *fait accompli* presented by the private companies unless an effective international authority was established soon.

[1] *Ibid.*, SR21, p. 107.

Eventually, some of the rich countries once more resorted to the strategy of 'if you can't beat them, join them' (a common ploy in modern UN politics). This affected, especially, attitudes towards the degree of international control necessary. As it had become apparent that there was a substantial majority in favour of some kind of permanent international 'machinery', developed countries, both western and eastern, slowly and reluctantly began to give grudging assent to this concept. But, as in the case of the 'common heritage', they sought in doing so to reinterpret the word to give it a rather different meaning from that which it had for developing countries. Even the Soviet and Bulgarian delegates, who passionately opposed any talk of 'machinery' in 1968, came to express cautious acceptance of the idea—but strictly according to their own definition. They remained opposed to any institutions which would have effective power over national states; and therefore to the main purpose of machinery in the eyes of developing countries. By equating 'machinery' instead with a general UN responsibility in the field of pollution and other matters, they both reduced some of their own isolation on the issue and removed part of the sting from the whole machinery idea. And it remained apparent, in the many dark warnings issued by western states against a complex institutional structure, or anything which could inhibit investment, that there were still many differing concepts of the type of machinery required.

This raised another problem: should the sea-bed be worked directly by an international agency? The rich countries heaped unlimited scorn on this proposal, asserting that it would mean that all the profits from development would be taken up in administration and that effective exploitation would be slowed down. It was denounced as a 'disastrous venture' (Belgium), 'premature' (Britain), 'Utopian' (Iceland), a proposal 'that would discourage sea-bed exploitation and narrow the benefits to the international community' (U.S.). The Communist states, which on *a priori* grounds might have been expected to be most favourable to this embryonic form of world socialism, were the most scathing of all about any such supranational proposal. The developing countries, on the other hand, though often accepting that it might not be immediately practical, contended that it was a possibility that 'should not be dismissed' (India). Kuwait stated definitely that 'the machinery should be empowered to undertake certain operational activities directly if the need arose'. And later (in 1971) some of the Latin-American countries came out with the proposal that a substantial part of exploitation should be undertaken directly by an international enterprise of this kind.

Then there was the question: could the system be imposed by majority vote? The rich countries, both of east and west, vehemently stressed the need for consensus and emphasized that a régime could only be set up by agreement; a shot across the bows of the more ardent among the developing, who might have hoped to use their voting power to push through a system quite unacceptable to them. Again, there were differences concerning the division of responsibility (between the international authority and the government receiving a concession) for supervision and inspection: the developed countries, especially Britain and France, sought to reserve a large role to the latter. The poor countries, conscious of the imbalance in techniques, laid great stress on the need for adequate training of their own nationals under any régime which might be established. They were insistent that any revenues obtained for international purposes should not be regarded as 'aid', and so reduce the commitments to aid of other sorts. Even *among* the developing there were differences between those which wanted the revenue to go direct to an international organization, or even the UN itself, and those which hoped it would go direct to governments.

By now the central points on which clear-cut differences of view were emerging began to become clear.

The first question was: should there be a limit to the area which any one state could acquire for exploitation? The United States claimed that there should be no limit at all. As a payment would anyway be made for international purposes on all production, the main objective should be to promote production of all kinds as rapidly as possible. This would both benefit the world economy and also maximize the royalties that would be paid. The fees to be levied on operators would still probably limit the amount any individual *company* held. As the U.S. representative put it:

At the present stage only a few nations had the ability or desire to exploit the resources of the area under discussion, and it was unlikely that all nations would be able to do so during the pioneer stage. If an attempt were made to restrict the activities of operators or nations to the areas to which they had exploitation rights, it would in effect discourage exploration and exploitation and reduce the revenue which might be shared by those unable to participate during the pioneer stage.[1]

The difficulties of finding an equitable means of distributing the resource potential were such that they must be regarded as one of the great disadvantages of any rule designed to limit the total area or the total resources to be held by any one party or state, or to allocate them in some way to nations. . . . Few existing national systems contained provisions limiting either the number of tracts or the total area which might be held by a given operator. . . . Such a policy would help to increase revenue during the period when the developing countries needed it most.[2]

But the United States was almost alone in this view. Even most of the other rich countries wanted some limit. France said specifically that no state or group of states should 'be able to put in a claim to more than a certain number of square-kilometres'. The British system, proposed about this time, explicitly favoured a limited allocation per state. And the developing countries were naturally still more concerned that a limit should be placed on the area which each state could hold.

The second point of difference was equally fundamental. This concerned whether the *initiative* in allocating blocks should be held by the international authority, or whether allocation should result from the applications made by states on the basis of exploration. Both systems had been used by national states in leasing their continental shelves. The United States foresaw the initiative being given to the companies or their governments, which would make their own bids for areas on the basis of exploration. The U.S. delegate held 'that the registration of claims or tracts on a "first-come-first-served" basis greatly encouraged operators, for it rewarded them with the exclusive right to exploit their discoveries. . . . Competitive bidding was generally regarded as the best means of selecting the most efficient operator'.[3] This was an understandable preference for a country such as the United States, with its highly developed sea-bed technology. But it was not a system that appealed much to many other countries. The British government proposed a system of direct allocation of selected areas of the sea-bed to governments, on a random basis, perhaps chosen by computer. The developing countries, where they expressed views at all, seemed to want the sea-bed authority to choose

[1] A/AC138/SR29, p. 39.
[2] *Ibid.*, SR29, pp. 29–30.
[3] *Ibid.*, SR28, p. 27.

particular areas of the sea-bed and divide these into blocks that would be distributed on a 'just' basis.

Next, should allocation be to states or direct to companies? There was at first a widespread opinion in favour of a three-tier system with allocation in the first place to states (it was, after all, the representatives of states who were considering the matter). Even most of the developing countries, which might have seen some advantage in allocation direct to companies (because it would have maximized the bargaining position of the international authority) tended to favour the three-tier system. Some seemed to see something sinister in the idea of allowing companies direct access. Others seemed to regard it as simpler to give a role to governments. The representatives of the United Arab Republic declared that licences should be granted 'at least initially to states, so as to avoid adding the complexities of international responsibility to the already difficult problems of an entirely new sphere of endeavour'.[1] The representative of Cameroon declared 'that concessions should be granted only to states, which would thus assume responsibility for all activities carried out by the governmental agencies or private firms authorised to operate under licence. . . . It would be easier to deal with a state or with regional or other internationally recognized organizations than with private corporations, which lack a clearly defined status in international law'.[2] Other countries, such as India and Kuwait, for example, felt that the alternative of allocation direct to companies should be retained at least as a possibility.

Finally, was there a need to take special action to avoid disruption of the markets of existing producers of the minerals? Both the United States and France argued that there should be no special measures of this sort at all. The problem should be tackled only in the general context of commodity arrangements: any restrictions would be applied to existing land sources as well as to those from the sea-bed. The U.S. delegate said that 'his delegation believed that that was a problem which should be tackled globally, without discrimination against ocean producers. To subject sea-bed production to special limitations would be a major deterrent to sea-bed exploration'. However economically rational such arguments, they could scarcely be expected to appeal to countries whose main source of income and foreign exchange

[1] A/AC138/SR34, p. 69.
[2] *Ibid.*, SR33, p. 49.

could be seriously threatened, or even cut off altogether, through sea-bed operations. Some developing countries, especially the producers themselves, were thus insistent on the need to ensure that existing revenues were adequately protected.

Thus though positions had drawn closer on some points, there remained deep differences on many questions, including some of the most fundamental. These were clearly revealed in the next phase, from the middle of 1970 onwards, when a series of national proposals began to be put forward.

National Proposals

By this time governments had had two or three years to consider the complex problems and to formulate their views about the kind of régime they would like to see. They had listened to the views expressed by other states. And they had reached the stage of being able to put forward comprehensive proposals, sometimes in the form of draft treaties, concerning the type of régime they hoped to see introduced.

In some ways, the most interesting and provocative was the first, put forward by the United States in August, 1970. This was based, in its delimitation of the areas where exploitation was to take place, on an entirely different principle from any other plan so far proposed.[1] Instead of a simple division between one zone (territorial sea and continental shelf) that was national and another that was international, the sea-bed was divided into three. There would be a fully national zone, ending at the 200-metre isobath; an intermediate area, to be called a 'trusteeship' zone, extending perhaps to the edge of the continental margin; and a fully international zone beyond that. The trusteeship zone would be in

[1] Somewhat similar proposals had however been put forward already within the United States. The American Assembly in May, 1968, proposed that the United States should not claim, and should urge others not to claim, permanent exclusive rights beyond the 200-metre line; that in adjacent areas beyond that depth the United States should issue licences for exploitation subject to the condition that they should eventually transfer them to an appropriate international régime. The U.S. Commission on Marine Science, Engineering, and Resources in *Our Nation and the Sea* (1969), p. 151, made a proposal for an intermediate zone, extending to the 2,500-metre isobath, or 100 nautical miles from the coast, whichever was the greater, in which the coastal state could determine who undertook exploration and exploitation, while in other respects the region would be an international zone.

principle a part of the international area and subject to certain international regulations. But within it the coastal state could act as 'trustee' for the international community, and undertake various functions and responsibilities there. These would include helping to ensure observation of the necessary rules concerning safety, pollution and so on. But the essential right which the coastal state would retain within this zone was that of determining to which operators licences would be issued. This would mean in practice that they would be issued mainly to operators of its own nationality (where they had the necessary technology). The coastal state would, moreover, not only have full discretion concerning whether, how, and to whom licences would be issued, but could also keep a proportion, perhaps up to a half, of the royalties it obtained from companies, together with any additional taxes it might seek to impose. Together, these would be an extremely important advantage for such a state. The continental margin not only represented altogether 20 per cent of the oceans' area; it was the part by far the richest in natural resources, containing, in particular, almost all the oil and natural gas of the sea-bed.

In the fully international zone, an International Sea-bed Resource Authority would issue licences against royalties, would undertake inspection, adjudicate disputes and make detailed regulations. The authority would include an Assembly, a Council, and an independent Tribunal. There would also be separate Commissions on rules, operations and the sea-bed boundary. The Council would have the main executive authority. Within it, the six most industrially advanced members would have reserved seats. Its decisions would require not only a majority of the whole Council, but also a majority of the six most developed members (in practice, 4 out of the 6 would need to agree). A kind of group veto for this class of country was therefore provided. Investments made before the treaty came into force would be protected (a provision which could give rise to a further race in pre-empting the sea-bed). Inspection would take place both in the fully international area and in the trusteeship area and would be undertaken by national states and the international authority. Apart from licensing fees and rentals, which would be modest, the main royalty payment would be a payment on production, based on the value of the minerals at the site. Revenues obtained by the authority would be used partly to promote research, development and technical assistance in sea-bed enterprises, and partly for international development.

Britain and France circulated more modest 'working papers' setting out their own proposals. They were in some ways alike, in that they provided for allocation of blocks to states, and a large role for national governments.

The British plan provided for the establishment of an international authority, which would be responsible for administering the region generally, with a plenary Conference of all members, and a Governing Body. The latter would be 'small in size in the interests of administrative efficiency'. It would be 'balanced to inspire confidence', and to reflect the interests of and the technical contribution which could be made by the developed and developing countries, land-locked and maritime. The authority would issue licences to member states. These would then be responsible for sub-licensing operators and ensuring that they maintained adequate standards and safeguards. The whole of the sea-bed outside national jurisdiction would be divided into blocks large enough to allow efficient exploitation, but 'small enough to allow fair opportunities to all states parties to the agreement'. The proportion open for application by each state would be laid down in the agreement, and if any state failed to ratify within an agreed period, arrangements would be made for its share to be reallocated (this could be a significant deterrent to any nation feeling that it might be better to go it alone). Some of the blocks would become open for allocation at intervals, and states would apply for individual blocks up to their own agreed proportion. Where only one state applied for a block, it would automatically get it; but where there was a dispute, allocation would be based either on the date of application, on mutual agreement between states, or on a random selection between them. Blocks would be allocated for successive periods of fifteen years, but there would be provision for relinquishment of parts at stipulated periods. Licensing fees would be paid to cover the costs of international administration; and royalties would be used for international development purposes. The international agreement would lay down general operating standards, whereas more detailed rules would be drawn up by the authority. The authority would undertake inspection to ensure that they were being implemented.

The French plan provided for two separate régimes: according to whether mobile equipment or fixed installations were being used. For the former, as in the dredging of nodules, only registration of economic activity would be required, and no exclusive rights would be granted. Certain rules of operation would be drawn up in an international Convention, which might also lay down the period of validity of each registration. For exploitation with fixed installations, as for oil or gas, states or groups of states would be allocated areas of the sea-bed for a given period, and would then themselves license companies to operate them. If there were rival applications for the same area, a decision would have to be made by the Conference of Plenipotentiaries, or some technical committee delegated by it. The general rules to be observed by states and companies in the area would be laid down in an international Convention. Allocation would be on the basis of the claims of states, but no state could claim a monopoly of the area adjacent to its coasts, and there would be a limit to the amount that any state could claim in any period of ten years. While the international authority would draw up general regulations governing exploitation and the protection of other users of the seas, each state could apply its own municipal law with respect to taxation, working conditions, social welfare, criminal law and customs, to the companies using its areas. There would be no payment of royalties or taxes direct to the international community. But the state holding each area could levy taxes on the companies operating there, and could contribute a part of this, if it chose, to any assistance programme—international, regional or even bilateral—of its choice.

The French and British plans thus both provided for a three-tier system, with national governments taking some responsibility for applying their own regulations to the operating companies. They both provided for a fixed ceiling on the number of blocks any state could hold at any one time. But the French plan provided for a quite separate system for dredging operations or anything without fixed installations, allowing virtually a free-for-all in this field. It accorded a larger role to the national state in regulating and enjoying benefits from its own area. Above all, it made contributions by the government to an international authority or for international purposes, voluntary rather than compulsory (reflecting, no doubt, the general French antipathy to international institutions).

The 'draft statute' for an international sea-bed authority presented by Tanzania in March, 1971, represented the first comprehensive statement of the views of a developing country on the régime. Part of the treaty was relatively uncontroversial. It provided for the creation of an international sea-bed authority, with an Assembly, a Council and a Secretariat. There would be a Distribution Agency responsible for recommending to the Assembly the equitable sharing of the authority's revenues; and a Stabilization Board with the power to fix prices of the raw materials sold, so as to 'balance the need of the world community for raw materials and the need for stability of the economies of producers of land minerals. . . .' It would issue licences for exploration and exploitation 'in accordance with such criteria as may be laid down by the Assembly'. The Council would have wide powers. It would determine the minimum and maximum areas in which rights would be granted, the payments to be made to the authority, performance requirements, rules, and standards concerning pollution. There would be no special representation in the Council; and voting would be by two-thirds majority. An interesting feature was that instead of distribution for development the income of the authority would be shared among all members in 'the inverse ratio of their respective contributions to the annual budget of the UN': a system which would mean that even rich countries would get their share of the revenues but most would go to those who needed it most. On the crucial question of the size of the international zone, it was laid down that each coastal state could choose whether this should be defined according to depth or to distance from the coast; this would ensure that any state which felt that it could be penalized by one or other basis could choose the most favourable to itself (this might reduce dissension on this crucial point).

The Soviet Union (in a document circulated to the Economic and Technical Committee entitled 'Proposals which may be studied in the Economic and Technical Committee'), at first favoured a very loose system of organization and did not mention machinery or institutions at all. Basically, national states should take responsibility for their own nationals but would 'co-operate' so far as possible, in their use of the sea-bed. But the draft treaty which the Soviet Union finally presented in August, 1970, showed that even the Russians had been edged marginally by the general consensus in favour of some system of international regulation. A good deal of the treaty was devoted to reaffirming traditional freedoms of the sea. Those undertaking 'industrial exploration and exploitation' of sea-bed resources should agree not to obstruct important sea-lanes, straits or 'points of intense fishing activities'; they should undertake adequate safety measures; they should give full notification of all installations to other nations and remove them when work was completed. States should 'co-operate' with one another to prevent pollution and contamination, damage to flora and fauna, to ensure safety, and over questions of research. States party to the treaty would ensure that all activities of persons 'under its jurisdiction or acting on its behalf' were conducted in accordance with the treaty', so putting the main responsibility on national governments. The treaty made no provision for any inspection agency. And its authors' dislike of supranational institutions was shown even more clearly in its insistence that 'none of the provisions of this treaty or the rights granted to the international sea-bed resources agency . . . shall mean that the agency has jurisdiction over the sea-bed and the subsoil thereof, or shall give the agency rights or legal grounds to consider the sea-bed and the subsoil thereof as owned, possessed or used by it, or at its disposal'. The Soviet Union was no doubt particularly opposed to the authority's acquiring jurisdiction over the sea-bed, as this would imply that no nation could undertake exploitation there if it did not become a party to the treaty.

Articles on many of the most crucial points were left blank, with only an indication in brackets of what the subject matter would be. This applied not only to the article on the limits of the sea-bed—this had been left blank in some of the other treaties—but also to those concerning the 'question of licences for industrial exploration and exploitation of sea-bed resources' and the 'question of distribution of benefits'. But the significant fact was that articles on these two subjects were provided for at all: for this marked a distinct advance on anything that Soviet delegates had said previously. It showed that at least the Soviet Union seemed now to accept the idea that there would be a system of licensing, and that payments might be made for international purposes. The treaty also provided for an International Sea-bed Resources Agency, with a Conference and an Executive Board. But in this, every member would in effect have a veto: it was laid down that 'decisions of the executive board on questions of substance shall be made by agreement', and majority voting would apply only to procedural questions. Moreover, it was provided that parties to the treaty 'may' become Members of the Agency; so clearly implying that they need not do so: nations might agree to co-operate and obey certain rules in the treaty without joining the international system of exploitation. The Executive Board of 30 members would meet at least once a year and would 'supervise the implementation of the provisions' of the treaty, and 'activities in connection with the industrial exploration and exploitation of the resources of the sea-bed and the subsoil thereof'. Its membership would be highly favourable to the Communist states. It was to be divided equally among six groups: the Socialist states (was China a socialist state?); the countries of Asia; of Africa; of Latin America; of all other areas, including western Europe and North America; and the land-locked.[1]

Another treaty was proposed by a group of Latin-American countries (Chile, Colombia, Ecuador, El Salvador, Guatemala, Guyana, Jamaica, Mexico, Panama, Peru, Trinidad and Tobago, Uruguay and Venezuela). The main feature of this was that it placed far greater emphasis than any of the others on direct exploitation by the authority. One of its constituent organs would be an International Sea-bed Enterprise to undertake 'all technical, industrial or commercial activities relating to the exploration of the area and exploitation of its resources'. The Enterprise could either undertake this work itself, or 'in joint ventures with juridical persons duly sponsored by states'. There was no provision for the division of the sea-bed into blocks to be distributed to states. Members would make contributions to finance the authority. There would also be a 'distribution among contracting parties of benefits from activities in the area', rather than a distribution for development to poor countries only. The Assembly, representing all members, would be the main policy-making body, able to decide on 'any matters within the scope of the Convention' and to 'give directions to the Council'. It would decide the criteria to be applied for contributions and distribution, which parts of the area would be opened for exploration and exploitation, and other matters. It would establish a Planning Commission to draw up plans for the development and use of the area and its resources. There would be a Council of 35 members, elected by the Assembly, which could take decisions by a two-thirds majority.

[1] There was not a monolithic view among the Communist States: Poland presented a working paper which differed from the Russian treaty in significant ways.

This would issue regulations on all activities in the area, supervise these activities, set rules and standards on pollution, and adopt technical assistance measures. The authority would establish oceanographic institutes for training, and would strictly control all scientific research. Moreover, as was to be expected in any proposal coming from the Latin Americans, the 'rights and legitimate interests of coastal states' (not defined) were to be respected in all activities.

But perhaps the most ambitious and interesting of all the treaties proposed by a developing country was that put forward by Malta. There were two guiding principles behind this proposal. One was that the entire area of ocean space represented a single interrelated ecosystem: it was no use devising the means of regulating one aspect—say, the exploitation of minerals—without regulating others at the same time—say, fishing. A single régime should be established, with a single authority exercising over-all control. The second basic principle, related to this, was that instead of a series of separate zones of jurisdiction for coastal states—territorial waters, contiguous zone, fishing zone and continental shelf—what was needed was one single zone for all purposes. 'The time had come' Dr Pardo said, 'to consolidate the multiplicity of limits of coastal state jurisdiction in ocean space into a clearly defined outer limit of national jurisdiction that recognised and satisfied the totality of the interests of the coastal state in the marine environment'. And the Maltese Treaty therefore proposed a single limit of jurisdiction at 200 miles. Within this the coastal state could reserve to its nationals all exploitation rights.

This seemed at first sight to align Malta, which had previously been outspoken against the extravagant claims of the Latin Americans and others, with the adherents of the widest limits possible. There were, no doubt, a number of plausible justifications for such a move. It could, as Dr Pardo himself claimed, 'avoid prolonged haggling' (by the wide-limit countries, yes, but not necessarily by the narrow-limit ones): by removing the opposition of those who were among the most fanatical in this respect, it could lead to more rapid agreement on a régime. It might even win the support of the Latin-American countries for proposals, such as Malta's, which were ambitious in other directions. And, incidentally, a wide limit could have been not entirely without advantages for Malta, itself an island state, surrounded by water, with known oil resources not far from its shores.

However, these first appearances were somewhat misleading. The reversal of attitude may have been partly a skilful tactical move. On closer examination, much of what Dr Pardo gave so generously with one hand he took away, more surreptitiously, with the other. For even within the broad national area that he proposed, governments would be subject to strict limitations on their sovereignty. They would be obliged to enter into various specific undertakings concerning freedom of navigation, including stronger 'rights of innocent passage', and on freedom of research. They would be subject to special responsibilities in the field of pollution. Above all, for all exploitation, even including fisheries, within part of the area at least, they would have to make payments to the international régime. Dr Pardo originally suggested in the Committee that, beyond 100 miles, the coastal state would have to begin making payments to the authority; 25 per cent of all its revenues for the first 50 miles beyond that point, 50 per cent for the next 25, and 75 per cent for the 25 miles beyond that (these percentages were based on total *revenues*, and not on a proportion of 'royalties', themselves fixed at an unknown level, as in the U.S. proposals, for example). In the Maltese draft treaty itself, however, no percentages were laid down, and the amounts to be paid were to be negotiated with the coastal state in a future Convention: a proposal which would, of course, have left the coastal states in a very strong position.

Thus, while in appearance Malta was appeasing the demands of the wide-limit countries for a 200-mile zone, it was in fact ensuring (at least in its original proposals) that the international zone—or, more accurately, the zone from which international revenues would be acquired—would begin at 100 miles. According to Dr Pardo's calculations, even from these semi-national areas alone (between 100 and 200 miles from the coast) 'the authority might within 5 years be acquiring $800 million a year, three-quarters of it from fisheries'. The Maltese proposal provided that the revenues should be used mainly to provide services in the marine environment—more research, monitoring stations, a network of scientific research establishments and marine parks—and training for developing countries. In the land-locked states, it could be used for the improvement of the environment there, preferably the protection of rivers and lakes.

On voting arrangements the Maltese proposals were more

realistic than those of other developing states. Because the central institutions would require powers of administration and regulation of a kind never previously granted to an international organization, they must be acceptable to the 'preponderance of world opinion measured in terms of properly represented population and power: the principle of one-nation-one-vote would have to be replaced by a system that would ensure an equitable balance between the interests of the different countries'.

Discussion on these various possibilities continued, in the Sea-bed Committee and its Sub-Committee I, for three years between 1970 and 1973. Eventually, a point was reached when an attempt could be made to collate a text for at least part of the treaty, showing the alternative formulations proposed. But the differences remained huge. The range of national proposals put forward represented widely divergent views of the type of régime that should be created, reflecting conflicting national interests. It would be no easy task to bridge the gap between them in the coming years.

The Conference Begins

These then, in brief, were some of the main problems which had arisen in relation to the régime. On some points there had been some small movement towards consensus. On many others there remained a wide gap. The most important questions to be decided were as follows.

(1) Should exploitation in the deep sea be undertaken directly by the international authority, or should there be a system of licensing?

(2) If the latter, should licences be granted to national governments for them to sub-license, or direct to the operators, whether private or public enterprises?

(3) On what criteria should licences be granted?

(4) Should the initiative in deciding which areas should be exploited lie with the operators, national governments, or the authority?

(5) Should licences be granted for limited purposes or resources, or for a general and unlimited right to exploit?

(6) What should be the limit (if any) to the area which any licensee, whether national government or individual operator, could hold?

(7) What should be the nature and scale of the royalties to be paid?

(8) Should royalties be payable to national governments as well as to the international authority?

(9) Should the revenues of the authority be used for specific purposes, such as sea-bed research or training, or given to development agencies, or handed direct to governments?

(10) If the latter, what should be the criteria on which distribution would be based?

(11) What type of powers would the authority require?

(12) Should the régime apply to sea-bed exploitation only, or to all aspects of ocean space activity, including fishing.

(13) What role should national governments play, for example, in securing compliance with the authority's regulations?

(14) How much of the system should be laid down in the treaty itself, and how much left to the institutions to decide later?

(15) What kind of voting system should be used within the authority?

(16) Should this provide any kind of veto for particular powers or groups of powers?

(17) What sanctions should the authority possess against violation of its regulations?

(18) Should the authority possess sovereignty in ocean space?

(19) Should there be regional régimes or systems, instead of, or as well as, the international system?

(20) What special rights should be accorded to coastal states concerning exploitation off their coasts?

(21) What measures should be taken, if any, to protect the traditional producers of sea-bed minerals?

(22) Should there be an obligation for the transfer of technology from rich countries to poor, and if so, of what kind?

(23) How would disputes concerning the system be settled?

(24) What would be the position of countries which refused to sign the treaty?

(25) Last, but by no means least, where should the boundary between the national and the international areas lie?

These were only the more important points of disagreement. Even if there had been some consensus concerning the basic type of system to be employed, therefore, there would still have been acute differences on many of these specific aspects of the régime.

To resolve them would be a huge and complex task likely to require many years of negotiation to complete.

The process of negotiation continued in the conference itself. The first session of the long-awaited Conference on the Law of the Sea opened in Caracas in June, 1974. To nobody's surprise it reached no final conclusions on anything. Inevitably, on the points still in dispute there were highly-publicized disagreements and these alone attracted the attention of the world's press. The reality however was that at the conference there was a perceptible advance on a number of points.

First, there was relatively rapid agreement on the vexed question of procedure. Already at the previous Assembly there had been reached a so-called 'gentleman's agreement', under which it was decided that decisions at the conference should be by consensus where possible. This was reaffirmed in the decision of the conference on voting. Only when agreement had ultimately proved quite impossible would there be recourse to a vote; and in this case a two-thirds majority of those present and voting would be required so long as that majority consisted of at least a simple majority of those participating in that session. Since all of those participating were in practice likely to take part in any important vote, nearly one hundred votes would be required for each major decision. It was also agreed that there should be a 'cooling-off period' for further thought and negotiation where necessary: if fifteen representatives wanted further discussion on any particular point, the vote would be deferred while further efforts were made to negotiate agreement.

Secondly, there emerged now a considerable measure of agreement about the main maritime boundaries (at least among the coastal states). Support for a twelve-mile territorial sea was now general. More significant, the concept of the 200-mile economic zone also won widespread support, among rich states as much as poor. This was not in itself surprising, since the concept was highly favourable to all coastal states: the acceptance of this principle would reserve a third of the oceans to such states (nearly half of them highly developed countries, such as the U.S., Canada, the Soviet Union, Australia and South Africa). But the fact that the major powers, such as the U.S., the Soviet Union, Britain, France, West Germany, and other important states (in addition to China, which had for long supported this concept) came out in favour meant that from this point it became overwhelmingly probable that such a zone would be incorporated in some form in the final solution.

But there were still differences about precisely what rights the coastal state would enjoy in the zone. The rich states complained that in the eyes of many Latin American countries the zone was being almost equated with the territorial sea: in other words, it was being used as a means to win them virtually full sovereignty in the area. By the rich countries, on the contrary, it was proposed that the coastal state would acquire there only carefully defined rights, primarily over economic resources. From this point therefore the effort became to define more exactly what *rights* the coastal state should have in the zone, and what *obligations* it would have there to respect the rights of other states, especially in relation to navigation. These obligations might include possible duties to land-locked states, for example to accord them fishing and exploitation rights, or to share a part of the revenues in the area with the international system.[1]

Some coastal states at the conference began to extend their claims even beyond 200 miles. Britain, Australia, Canada and others claimed economic rights beyond that distance, to the edge of the continental margin (in other words to depths of 3,000 metres or more). This appeared to be based on the International Court's 1969 judgement concerning the 'prolongation' of the land mass. It was in fact a considerable stretching of that judgement: since the only clear limit to the prolongation of the land area is that which occurs at the edge of the *shelf* (usually below 200 metres in depth), that judgement seemed to imply that this should be the limit of jurisdiction: at most this should be the bottom of the slope.

There remained sharp differences on other points. On the question of passage through straits, especially as it affected naval vessels, there was still a conflict of views, though it was gradually narrowed down. On exploitation in the deep sea area, there were big differences, especially on the principles to be applied by the sea-bed authority in granting licences: the poor countries wanted it to have total discretion on this point, the U.S. wished licences to be granted almost automatically to any qualified enterprise,

[1] Many countries of Latin America and Africa accepted the idea that land-locked states could share in fishing activities in the zone of neighbouring coastal states—an inexpensive offer since the land-locked states had neither fishermen, fishing vessels, or ports—but rejected their right to share in exploitation of mineral resources there.

while the EEC proposed that a single enterprise should have the right only to six sites of a maximum of 50,000 square kilometers each. This was a vital point, which could determine to a considerable extent the distribution of benefits from the deep sea area and which could become a main centre of dispute.

The second session of the conference took place in Geneva in March–May 1975. Again the conference was hailed as a failure, merely because no final agreement was secured. But in fact on a number of points substantial progress was made once more.

'Free transit' through straits, including overflight and submerged passage by submarines, to which the super powers attached importance became more widely accepted. There was increasing acknowledgement that the coastal state should accept duties as well as rights in the economic zone, including a duty to accept international pollution standards, and a willingness to accord unimpeded navigation rights. There was more agreement on the need for a binding procedure for settling disputes. Finally, on the transfer of technology, there was increasing recognition, even among the rich states, that there was need for an obligation to undertake this to be spelt out; and that the authority itself should have an important task in arranging it, both through its licensing arrangements and in its own activities.

Even on the central question of exploitation in the international area there was some advance. There was increasing discussion of a compromise between the extreme view that all exploitation should be undertaken directly by the authority (or rather its Enterprise), and the opposite, U.S., view that private firms should enjoy a largely free hand (apart from the duty to pay royalties on their profits). Discussion increasingly focused on the concept of 'joint ventures', arrangements entered into between the authority (or the Enterprise) and private or state enterprises. An alternative proposal was the so-called banking system: an exploiting company should propose exploitation in two separate blocks of roughly equal value – the authority would then allocate one to the company and another to a developing country for international exploitation. This would reduce the advantage of the 'initiative' to developed states. The authority could be sure that it controlled sites of some value.

One point on which there remained considerable conflict was the question of scientific research: how far did a coastal state have the right to control *all* research within its economic zone (as many poor coastal countries claimed) or only research that was resource-orientated (as the rich countries demanded, quoting especially the 1958 Convention on the Continental Shelf, which laid down there should be freedom of research in the continental shelf). Here too a compromise was proposed. A group of four developing states suggested that any group or person applying to do research in the economic zone of another state would indicate whether the research was fundamental or resource-orientated. If the research was related to resources, the coastal state could refuse permission. In other cases it could indicate whether it wished to participate itself, or receive the results. If the coastal state made no reply, research would go ahead. It should not 'normally' prevent pure research.

As important as the progress on particular points was agreement on a single negotiating text on all the three main areas under discussion (the regime; maritime law, including limits; pollution, research and transfer of technology). Texts were eventually formulated by the chairman of each committee, with substantial assistance from the UN Secretariat, and reflected the outcome of discussions so far. They committed no governments, but they could serve as the basis for subsequent negotiation and an indication of a possible solution in each area.

The single negotiating text (SNT) provided for the establishment of a 200-mile exclusive economic zone. Within this zone the coastal state would enjoy the right to exploit sea-bed resources, exclusive jurisdiction over fishing and other economic activities, and 'jurisdiction' (not necessarily exclusive) for the purpose of preserving the marine environment.[1] In relation to fishing, the coastal state would be obliged to promote optimum exploitation, and so to allow access for the fishermen of other states to the extent necessary to achieve the allowable catch. At the same time, it would be allowed, and indeed obliged, to take steps to conserve stocks. Landlocked states were to be given the right to participate in exploiting fishing resources in these zones 'on an equitable basis' (whatever that meant). For anadromous species the states in whose fresh waters the fish originated were given regulatory powers.

So far as other maritime boundaries were concerned, the text provided for a 12-mile territorial sea, within which the right of

[1] The coastal state would also be given rights, under the proposed text, for mineral exploitation even beyond the 200-mile limit, to the outer edge of the continental margin: within that area, however, it was suggested that some royalties should be paid to the international community.

innocent passage was to be protected: this was defined as passage which 'is not prejudicial to the peace, good order or security of the coastal state'. Within fully international straits, that is, straits between two parts of the high seas, the coastal states would be obliged to accord free transit, and its right to make regulations applying to shipping there would be limited. In straits leading only to the territorial sea of another state (such as the Straits of Tiran), the coastal states would only need to accord the right of innocent passage, where its right to impose regulations would be less restricted. Within the territorial sea the coastal state would enjoy certain powers to regulate pollution and navigation. For archipelagoes the text incorporated the idea of 'archipelagic waters', as proposed by Indonesia and other states: in such waters a state would enjoy some rights to control navigation, but would not have all the rights of the territorial sea or the economic zone.

Over pollution a state would have the 'obligation to protect and preserve all the marine environment'. States should formulate national regulations on land-based pollution and all states should seek to establish regional and global standards to this end. It was recognized, however, by most (though not by all) states that for ship-based pollution there could not be a variety of national standards (for example on the construction of ships). So here it was necessary to establish 'through the competent international organization or by general diplomatic conference . . . international rules and standards for the prevention, reduction and control of pollution of the marine environment from vessels'. The main responsibility for enforcing these rules would be with the flag-state. But the coastal state would also be given powers to inspect and arrest vessels for violations, and, if the flag-state did not take proceedings, to take judicial action against them. Similarly, the port state (the state at whose port the vessel subsequently called) could take action against vessels to ensure compliance with pollution standards. In areas where pollution might cause 'irreversible disturbance of the economic balance' coastal states could establish their own laws and regulations to protect the marine environment.

Most of these represented reasonable compromises, which might well be accepted in the final text. The most controversial points in the text concerned exploitation in the international area. Here the text largely reflected the view of developing countries: that is, it favoured a tightly-controlled international system. An International Sea-bed Authority was to be set up, which would include, besides an assembly, a council and a secretariat, an economic commission, a technical commission, an Enterprise which might itself undertake exploitation and a tribunal to settle disputes. Normally operations in the zone would be undertaken by the authority, but provision would be made for joint ventures with private corporations, or for service contracts. The benefits of sea-bed operations would be equitably shared and the authority would promote the transfer of technology to poor countries. The authority would also be obliged to protect the interests of states already producing the minerals contained in manganese nodules. In most respects these proposals reflected the views and interests of poorer countries with little sea-bed technology themselves.

The proposed system, however, giving almost unlimited power to the international authority in the international area, was by no means acceptable to the developed countries and might, in their view, have delayed development for many years, since the authority itself would have no expertise. In the spring of 1976, therefore, after new negotiations, a 'revised single negotiating text' (RSNT) was produced, which represented a compromise position. This would have allowed a 'parallel system': with half the area to be developed by private companies under licence to the authority in return for a royalty, and the other half by the international Enterprise. At a later session that year, however, in New York, this system was in turn challenged by some developing countries, on the grounds that it gave too much advantage to the rich countries from which the private companies came; and because the Enterprise might lack both the finance and the technology to compete on equal terms. To meet the latter point, Dr. Kissinger, on behalf of the U.S., offered financial and technical assistance to the Enterprise so that it would be in a position to begin operations on almost the same time-scale as private companies. He also offered a review of the entire system after 20 or 25 years, so that the developing states could feel they might, in the future, when their technical level was more advanced, be enabled to play a bigger role.

In the early part of 1977 it seemed possible that an agreement might be in sight along these lines. To satisfy developing countries such an agreement had to show that they would have an effective role, not only in taking part in the authority but in operations: a share in the action itself. This would involve showing them that the Enterprise would be made financially and technically viable, either through support from governments or through joint ventures with private companies. They would also need to be shown that the system was not necessarily permanent, and would be reviewed after a certain period in the light of changed conditions (the difficult question concerned what would happen if there was no agreement at that point). To satisfy developed states the system would require to ensure that the valuable and badly needed resources of the sea-bed would indeed be exploited and within a reasonable time-scale; and, since only a small number of Western private companies at present possessed the necessary technology, this meant that such companies should be given a reasonable opportunity to take part in this work. It was not impossible that a compromise on these lines, allowing a role both for companies at the Enterprise, might secure the balance of interests that all had been seeking.

There was no certainty that any text of this kind would receive final endorsement. In some cases individual powers reserved their position. But in many cases the text did represent reasonable compromises between what had initially been deeply divided positions. It began to look as if a final solution (if there could be one at all) would be not too far away from that contained in the RSNT.

There was however one group of states whose views had been little reflected in the texts produced, if only because they were continually in a minority. These were the land-locked and other 'geographically disadvantaged' states. For them the reservation of the most valuable parts of the sea-bed, including everything within 200 miles from the coast and perhaps some parts even further, to the coastal states, not only reduced the revenues which would be available to the international régime, but largely excluded them from any direct participation in exploitation of those areas. These states by now numbered 48. This was just about a third of the total, a block sufficient to prevent any convention being agreed without them. The group had, for various reasons, until this point, failed to maximize the bargaining power this strength accorded to them. It seemed likely that from this point they might use their bargaining power to greater effect: for example to ensure that there was an obligation on coastal states to share revenues with the international régime not only in the continental margin beyond 200 miles, but within the 200-mile economic zone itself (as the U.S. had already proposed).[1] A solution on these lines, by accommodating the states that were already geographically disadvantaged, and redressing to some extent the gains which the coastal states were securing for themselves, might, perhaps, better than any other, secure the long-term balance of interests among states which could alone provide the final solution.

Divergent views on such points, as expressed during the negotiations, reflected deep underlying differences of interest, both among individual nations and the groups to which they belonged. And it is to these conflicts of interest that, in considering what are the prospects of reaching final agreement on the sea-bed régime to be established, we should now turn.

[1] A distinguished study group of the Trilateral Commission, representing influential figures among the most highly developed states of the world, during 1975 recommended a solution which included revenue-sharing in the economic zone of this kind.

The Future

CHAPTER 11

Basic Interests and Issues

Interests

Governments exist—or believe they exist—to protect *national* interests. They recognize, in a theoretical way, the need for solutions which promote the wider interests of all mankind. But they assume that such solutions can only be the end-product of a fight: of bitter and protracted wrangling among many different representatives, each bargaining for their own nation, rather than from a common, disinterested effort among all to devise the system that would best serve the interests and promote the welfare of the world as a whole. They accept the need for mutually acceptable arrangements. But their first aim is to ensure that the arrangements are acceptable to themselves. This is the inevitable effect of a situation in which political power and political parties are still organized on a national rather than an international basis.

It was perhaps inevitable, therefore, that the policies of virtually all the governments involved in the sea-bed negotiations have been directed to promoting the immediate interests of their own state, calculated in the most cynical and self-seeking terms, in the confident (and not unjustified) expectation that other governments would undoubtedly do the same for theirs. Because any régime to be agreed must satisfy at least partially the interests of the chief states and groups taking part in the discussions, it may be well at this stage to describe the basic interests which have been at issue for them and to consider how they have affected their attitudes on particular questions.

The interests of the United States are of a complex kind. The first, and perhaps the most important, is the fact that it is the power by far and away the most advanced in undersea technology. Hundreds of millions of U.S. dollars have been spent in recent years, both by the U.S. Government and by U.S. companies, in developing capabilities in this field. The U.S. government alone at present spends over $500 million a year on marine science and technology, while U.S. industry is spending at least ten times that amount. U.S. oil companies already have long experience of off-shore working, and are improving their underwater technology by rapid bounds. In addition, a number of U.S. companies, as we saw in Chapter 1, have become actively involved in exploring the possibilities of the commercial dredging of nodules. Within the Government, the National Council of Marine Resources, Engineering and Development, placed directly in the executive of the President, and with the Vice-President as its nominal chairman, together with the advisory Commission on Marine Science, Engineering and Resources, and a network of committees beneath this, give support to marine activities and research and help to formulate government policy.

This huge technical lead meant that, whatever régime was finally chosen, U.S. companies were certain to play a major part in the actual exploitation work. But the nature of the régime could have a big influence on the *conditions* under which this took place. Probably the optimum régime for U.S. *companies* would be one providing the maximum possible freedom in the deep sea, but also ensuring security and stability for any exploitation undertaken: perhaps a system of registration alone, or even a complete absence of control (for in practice U.S. exploiters, in competing for sites, could ultimately expect the protection of the U.S. Navy).

For the U.S. government, interests are somewhat different. Clearly it shares, and actively promotes, the interests of U.S. companies. But a régime of total freedom might provide no revenues for itself. It could create a danger of severe pollution and other problems through inadequate control. And it could lead to serious international frictions, resulting from an uncontrolled scramble, both among companies and among nations, for the most favourable deposits. Indeed, this last danger, the instability of the conditions provided, might in practice seriously qualify the benefits of a free system even for U.S. companies.

The interests of both the U.S. government and U.S. companies are affected by a second factor, unrelated to the level of technology.

This is the interest of a nation with an unusually long coastline in two oceans. This gives it a continental margin extremely large in area and, in parts at least, also extremely rich in resources. Part adjoins large-scale existing deposits of oil and gas in Texas and California. Part holds some of the only known relatively shallow deposits of nodules (on the Blake Plateau). Thus the United States shares, for quite different reasons, the interest of some Latin-American countries in securing a considerable degree of control for coastal powers over exploitation of wide areas off their own coast, even beyond the continental shelf.[1]

Finally, however, the United States has further interests, which in some ways might have conflicted with those relating to sea-bed exploitation. In financial terms it is much smaller but in political terms highly significant. One is as a naval power, interested in checking Soviet naval strength. The United States is determined to maintain the maximum freedom of the seas, especially free navigation through straits. The second is as a large-scale fishing power, with highly mechanized fleets, which could acquire considerable benefit through the enjoyment of free fishing rights off the coasts of other states, beyond territorial waters but within the continental shelf area. A general economic zone for the coastal state, as some have proposed, would inhibit this freedom. This meant that it was important for the United States, as for other fishing and maritime powers, to emphasize the clear distinction between rights on the sea-bed and those on the surface of the waters. And it was particularly important to resist a creeping jurisdiction, under which coastal states, on the pretext perhaps of acquiring control of exploitation or pollution, in fact acquired a general and unlimited jurisdiction many miles from their own coasts.

All of these interests were brilliantly harmonized by the United States in its proposal for a 'trusteeship zone'. This assured a narrow limit of jurisdiction for fishing, security and most other purposes; but at the same time allowed coastal powers to acquire substantial control over large areas of the sea-bed beyond that

[1] The U.S. hard minerals industry, which was interested in nodule production, was more concerned about the nature of the régime, than about the position of the boundary. The U.S. Navy favoured a *narrow* boundary. For an interesting discussion of the struggle between conflicting U.S. interests, see A. D. Hollick, 'Seabeds Make Strange Politics' in *Foreign Policy*, Winter 1972–3, p. 148.

area for exploitation purposes, to the benefit of both technologically advanced companies and the government of the coastal state. It satisfied the strong environmental lobby within the United States by providing a comprehensive and effective system of pollution control, but did not thereby give up valuable economic rights which the United States wished to retain, within the continental shelf and beyond. It was truly internationalist by offering the régime revenues from the 200-metre line; but secured at the same time even larger revenues for the coastal state from that area (including taxes on its own companies). In other words, it maximized most of the economic benefits which a country with a long coastline, such as the United States, might acquire from its continental margin area, without at the same time requiring concessions to the Latin-American countries in relation to fishing rights or threats to U.S. naval power. Other policies of the United States within the Committee—the attempt to prevent any limit on the holding for each state, to secure a system of passive licensing of areas applied for, rather than active allocation by the sea-bed authority, to maintain what was virtually a veto for the most developed countries within the authority, and other proposals—complemented the trusteeship idea in promoting U.S. interests but also providing large international revenues.

The interests of the Soviet Union are in some respects similar to those of the United States. Far more than the United States, it is fearful of according too much authority or control to any international body. It wants maximum freedom of movement for its navy. Again, it is against a system involving too much redistribution. The Soviet Union could perhaps finally become advanced in sea-bed technology, although it began working the sea-bed only in 1969. Unlike the United States, it will probably not, however, enjoy rich resources within its own continental margin, such as it is, and therefore is even more anxious to secure full access to other parts of the sea-bed.[1] For all these reasons, the Soviet Union had every reason to favour the maximum possible freedom of exploitation and the minimum possible international control.

The Soviet Union therefore, as we saw, for long favoured a system merely of 'co-operation' among states rather than any

[1] According to *The New York Times* of April 24, 1971, a conference of Soviet geologists at Riga at that time decided on a programme of exploration in the Atlantic and Indian oceans to select sites for exploitation.

attempt at international *control* of sea-bed activities. It persistently denied that there was any need for haste in formulating a régime. It insisted particularly strongly, as a member of a small minority group, on the need for consensus: for a 'universal' agreement on a régime 'which could be reached on the basis of the principle of sovereign equality'. It denied that common heritage meant 'common property' and preferred the 'more realistic' interpretation that it meant that the sea-bed 'shall be used jointly by all states without any discrimination whatever'.[1] And it attacked proposals for 'the participation of the international community in the proceeds and benefits of the exploitation of the resources of the sea-bed' as 'premature'.[2]

At the same time, as a large international fishing power like the United States, as well as a naval power rapidly extending its activities and influence in many oceans of the world, the Soviet Union was for a long time opposed to any attempt to extend national jurisdiction on the surface of the waters, whether for fishing or for any other purpose, beyond twelve miles. It therefore consistently took the line that questions of jurisdiction, at least so far as the surface of the waters was concerned, had been finally settled in the 1958 Geneva Conventions, and should not on any account be reopened at the present stage. It was even more scathing than developed western countries at the increasing tendency of 'certain states', meaning the Latin Americans, to extend their territorial waters and other rights far out into the oceans. It pointed out that, by adopting the 200-mile limit for its territorial waters in the Atlantic, one Latin-American state had 'extended its jurisdiction over an area of the high seas which exceeded 1.5 million square kilometres—the whole of the territories of France, Britain, Italy, Spain, Belgium and the Netherlands put together', while another had 'extended its jurisdiction over 3 million square kilometres'. And it pointed out correctly that if this were generally done, including the extension of jurisdiction from islands, it would take about half the ocean and it would become 'pointless to establish a régime at all, since it would in practice apply only to the ocean depths and other areas of the sea-bed which were virtually inaccessible for exploitation'.[3]

[1] A/AC138/SR30, p. 15.

[2] *Ibid.*

[3] A/AC138/SR56, p. 150. The Soviet Union stated quite baldly in 1972 that its agreement to a treaty according rights in the sea-bed would be conditional

Both these attitudes, the demand for a minimal régime, and resistance to any attempt to extend the jurisdiction of coastal states, inevitably brought the Soviet Union into increasing conflict with the developing countries as a whole in the former case, and the Latin Americans in the latter. After a period of peaceful co-existence, in which both groups still seemed to be bidding for the support of the other, these differences of interest came more and more into the open. By March, 1971, the Soviet delegate was openly attacking the attempts 'of some delegations' which were 'obviously trying to give the machinery excessive powers and to transform it into a supranational organ which would direct the activities of states on the sea-bed. Obviously they imagined that the machinery, and the immense powers they proposed to grant to it, would be controlled by them and used in their interests. That approach . . . was unrealistic. In present-day conditions, when states insisted on sovereign rights, the international community could create organization only to coordinate the activities of states and not to direct them'.[1] On the question of limits, too, Soviet and other Communist delegates increasingly came out into the open against the Latin Americans. They attempted to divide them from other developing states by demonstrating the divergence of their interests. The Polish delegate, for example, declared: 'Some representatives had displayed great eloquence in their advocacy of their countries' interests, but in so doing had presented a rather one-sided and oversimplified picture. . . . By extending the territorial sea and the exclusive fishing zone, not all and not only developing countries would be favoured, but only countries with rich offshore fishing grounds, whether developing or developed'.[2] Such conflicts sharpened as time went on.

The interests of Britain and France are in an intermediate category. Both, especially in the case of Britain, have large areas of continental margin, potentially very rich in resources. Both might eventually have a not inconsiderable underwater technology, though they remain, and will always be, far behind the United States. Both are maritime powers with a traditional interest in the freedom of the seas.

---

on agreement on a 12-mile territorial sea and a satisfactory accommodation in fishing rights.

[1] A/AC138/SR56, p. 155.

[2] A/AC138/SR54, p. 119.

But their position is by no means identical. France has had a strongly nationalistic government and a considerable distrust and distaste for international organizations. It has, partly through the pioneering efforts of Piccard and Cousteau, a more developed oceanological capacity, at least in the scientific fields. Finally, it has strong strategic and other interests in the Mediterranean, which it was reluctant to see endangered by an international régime established there. These positions were reflected in its own proposals. France demanded, more than any other country, a dominant role for national governments in the management of the régime. In its proposal that the dredging of nodules should be freed from all control other than registration, France reserved the possibility that its own industry could play a significant role in such activity all over the oceans. And, by stressing, with Italy, the importance of allowing separate régimes for enclosed or marginal seas, it reserved the possibility of establishing some special system within the Mediterranean which could protect the interests of the coastal states there.

Britain's position was somewhat different. It not only already possessed a very large shelf, but had the prospect of a still richer continental margin beyond (especially if it could secure recognition of Rockall as part of the United Kingdom). Because of the activities of British oil companies all over the world, especially in the North Sea, its oil-drilling technology is probably more advanced than that of most other western countries. On the other hand, British political attitudes, at least under the Labour Government, were considerably more internationalist than those of France. These positions conditioned British policies. The British representative stated quite frankly that his government was inclined to favour 'a reasonably deep limit to national jurisdiction, or a combination of a depth and width to give a broad deep limit'.[1] Britain favoured a system under which international companies of advanced technology might be in a position to undertake exploitation on behalf of any government, on terms which they could negotiate for themselves. But at the same time it supported a distribution of the sea-bed among states, and of the international revenues derived from it, which could be relatively favourable to developing countries including the land-locked

[1] A/AC138/SR39, p. 20. After initial doubts, Britain later gave some support to the U.S. trusteeship proposal, which would likewise have assured it control over a wide margin.

(though its value to non-coastal states would be sharply reduced by the wide shelf limit that Britain favoured).

Most of the other developed western states support positions not unlike those of Britain and France. They tend to favour licensing to governments alone; to oppose direct licensing to companies; and, especially, to oppose exploitation by an international organization. They want a considerable role for national governments in controlling and supervising operations. They want a system which would promote investment and provide the necessary 'stability' and 'confidence'. Their positions on the delimitation of the area vary (if only because, except for the Mediterranean countries, most had no continental margins beyond the shelf anyway). Most would have been prepared to see a larger proportion going to the international zone than Britain. Sweden, Belgium and the Netherlands (the latter of which are both shelf-locked) are inclined to support fully internationalist solutions. Spain takes a fully Latin-American line.

The developing countries have directly opposite interests. Being deficient in technology and capital resources, they would have been at an overwhelming disadvantage in any competitive scramble for deep-sea resources. Being have-nots, they have a general interest in a strong régime, with a powerful central authority which would strictly control exploitation and ensure the maximum redistribution of revenues. Because their voting-power is strong, they have an interest too in a system in which as many decisions as possible are vested in the proposed central sea-bed authority, which could be expected to preserve and promote their interests, and in opposing any special voting system by which the developed are given special voting privileges. Because they wish to reserve their options if no satisfactory régime can be agreed, they are mostly not prepared to make any undertaking concerning sea-bed limits until the details of the régime have been established. And because they are conscious of their own deficiencies in undersea technology, they are unanimous in demanding a substantial programme for assistance to developing countries, both within and without the régime, in this field.

In other ways, however, their interests differ radically, according to geographical situation and other factors. As the representative of Singapore pointed out, there were at least nine categories of interests among the nations of the world in respect of the sea-bed: the coastal states with extensive continental shelves; the coastal states with little or no continental shelf; the land-locked states; the shelf-locked states; the coastal states dependent on fishing in home waters; the states whose economies were dependent on liquid or hard minerals found in the sea-bed or likely to be found there; the archipelago states; the developed maritime states; and maritime states possessing an underwater technology.[1] In fact this list could be extended: there are the states dependent on distant-water fishing, both developed and developing; there are countries with especially small or large coastlines; there are the countries that are large-scale importers of sea-bed products and that which might have an interest in the most rapid and cheapest exploitation; and there are any number of combinations amongst these various factors.

The Latin-American states have consistently sought to obscure any such differences of interest among the developing nations. They represented themselves, on every possible occasion, as fighting a gallant battle on behalf of the poor countries as a whole against the depredations of 'imperialists'. They even managed to assert from time to time that, in claiming wide limits to protect their interests, they were defending the principle of the 'common heritage of mankind': a feat of reasoning that provoked some barbed comment from some of the land-locked countries, such as Nepal and Afghanistan. These not unreasonably pointed out that such claims, far from assisting poor countries generally, reduced international redistribution and so deprived the geographically underprivileged of part of their 'heritage'. Indeed, as the representative of Nepal pointed out, 'the policies and views of a land-locked developing country such as Nepal might in many respects be closer to those of a land-locked developed country than to those of a developing country with direct access to the sea'.[2]

China has sought to present itself as a leader of developing states against super-power hegemony; many of its interventions have been devoted to somewhat intemperate attacks on the Soviet Union. In reality, however, as a state with a long coastline, China has taken an extreme coastal state position. It has supported the claim to a 200-mile territorial sea of Latin-American states, though it has not yet claimed this for itself. It has demanded a wide jurisdiction, for economic, pollution and other purposes. China wants the 'international management of all mineral and

[1] A/AC138/SR60, p. 27.
[2] A/AC138/SR55, p. 135.

living resources in the international area'. And it wants all the Geneva agreements of 1958 to be scrapped and replaced.

The most fundamental difference of interest of all, therefore, was not that between rich and poor. It was that between coastal countries—especially countries with long coastlines (for a considerable proportion of coastal countries had only very small coastlines, and therefore small potential sea-bed resources)—and the rest. The former could hope that, through wide limits, they might appropriate substantial resources, living or non-living, in the oceans off their shores. The non-coastal, on the other hand, must automatically lose under such a system, since the sea-bed resources so gained must be subtracted from those available to the international régime, which alone could benefit non-coastal states. It was this basic division between coastal and non-coastal which began to become more and more explicit as the discussions proceeded.

The Latin Americans, led by Peru, Chile and Ecuador, made themselves the spokesmen of the extreme coastal state position. There was indeed considerable irony about this fact. For their basic concern, as we saw, was almost entirely about fisheries. That interest did not in itself require wide claims to continental shelves, still less to territorial waters. It could have been better satisfied merely by demanding wide preferential fishing zones for such states. On the contrary, they might have had the most to gain from *narrow* boundaries and the largest possible international area, as, with narrow shelves, they had virtually no sea-bed resources of their own. The wide limits they demanded would give them little or nothing in the way of mineral exploitation but, by impoverishing the international system, would reduce the revenues that they might acquire through redistribution. They were so dominated, however, by their concern over fishing rights, and the running battle with the United States which stemmed from it, that that issue alone dictated their policy. They believed that their demands for wide fishing rights were in some way strengthened and legitimized in being accompanied by quite irrelevant claims to jurisdiction in general; and they were thus in effect prepared to damage their more basic interests in relation to the sea-bed itself. Only the Caribbean states, with Mexico, Colombia and Venezuela, recognizing this distinction, proposed a twelve-mile territorial sea together with a 200-mile economic zone, or 'patrimonial sea', a lead later followed by other developing countries.

The other special group among the developing nations are the land-locked countries (together with the shelf-locked, which had almost identical interests). These obviously have an interest in the maximum possible internationalization, and the maximum possible redistribution. This means that they must be against a wide national area; against any other special treatment of the coastal state, as in the U.S. trusteeship proposal; and in favour of the highest possible royalties for international purposes. They are also likely to favour a plan, such as that of the British, under which every nation regardless of its situation could acquire its own plot of sea-bed. In this way they might acquire a direct interest in the sea-bed which they could never otherwise enjoy. As the discussions proceeded, they became better organized, making common policies and initiatives as a group.

The Key Conflicts of Interest

There are thus many variations in the position of *individual* countries, both in accordance with interests and according to how far they favoured a more extreme or a relatively moderate solution (for example on voting rights, the distribution of royalties, or the powers of the international authority).

But there are two basic divisions. The developing countries have in common the underlying interest of have-not powers. This closely mirrors the political stance of the underprivileged within states. Just as the latter seek to remedy their position by the establishment of a strong state authority or even outright nationalization of resources, to effect redistribution from rich to poor, so the underprivileged nations now demand a strong sea-bed authority, and internationalization of resources, to achieve redistribution—of wealth, technology, and above all control—from rich nations to poor.

Yet it is arguable that even more important than that between rich countries and poor is the division of interest on the boundary. For this will determine who gets the benefits of many of the most valuable and easily accessible sea-bed resources. If a wider outer limit is agreed, the area available for international exploitation will be reduced, and the bulk of the accessible resources will go to the coastal states, especially the large ones. The areas closest to the coast will anyway probably be those richest in resources. They certainly will contain most of the oil and gas of the sea-bed.

If the boundary is pushed out to 2,500 metres in depth, for example, as has once or twice been suggested, or to the 200-mile limit now being widely proposed, this would mean that by far the most valuable part of the area is appropriated by the coastal governments for their own benefit. The countries fortunate enough to have long coastlines and productive continental margins would become the chief beneficiaries (including perhaps a few tiny islands, acquiring large windfall profits from the wide shelves they controlled). Countries with small coastlines, the land-locked and the shelf-locked (and, to some extent, the developing as a whole) would be deprived of the benefits of resources already acknowledged to be the common heritage of mankind.

One of the basic questions which arose here was whether the boundary should be calculated on the basis of depth or of distance. There was an increasing body of opinion which agreed that a limit by depth alone was unjust and irrational. Some countries were unduly favoured, with wide and shallow shelves, whereas others possessed virtually nothing at all. A limit based on distance would at least give all countries shares roughly proportionate to their coastline. One based on depth gave entirely unpredictable sea-bed areas. If the boundary was to be widened, a limit by depth became even more absurd, as it would bear less relationship than ever to the physical shelf (which on average ended at around 150 metres in depth), and so would have virtually no rational justification. There was thus an increasing demand for a formula either based on distance alone, or containing some combination of depth and distance.

The countries with narrow shelves, such as those on the west coast of Latin America, inevitably especially favoured a definition by distance. Those, like Britain and the United States, with gently sloping shelves, might have preferred an extension of the depth limit, say to 500 metres or more; the edge of the continental margin, at 3,000 metres or more, was even occasionally mentioned, officially or unofficially. But as time went on, there was an increasing disposition to think in terms of a new area of jurisdiction, an economic zone that would be measured in distance from the coast.

However, whatever the basis used, the fundamental question was still the same: wide boundary or narrow? Here interests were not only conflicting, but also difficult to calculate. For coastal developing countries, the gains which they might make through international redistribution had to be weighed against the possible gains they could make from their own continental margin if it came under their own control. Countries with a long coastline, and so with a potentially large shelf, such as Brazil or India, had a quite different interest from one with only a short coast or a steeply sloping shelf, such as Lebanon or Ecuador. Even this calculation was further complicated by varying interests in different activities of the sea-bed. The Latin-American countries and Iceland had to weigh the advantage of a fully international régime in the field of mineral resources against the possible benefits of keeping a large area under their control for the purposes of fishing. Countries such as the Soviet Union and Japan had to weigh the advantages of the fullest possible freedom of action in relation to fishing against the possible benefits of extensive areas under national control for the purposes of exploitation.

The interests of developed countries were complicated by other factors. For some, such as Britain and perhaps France, a wide limit for the national area seemed to be advantageous, as they themselves had broad and particularly productive continental margins. But this would have meant in turn that their own companies would probably have been excluded altogether from large continental margins belonging to other states. Some rich countries might calculate that they would do better if opportunities for their own companies in the international area were maximized, and the area reserved to coastal states kept to the minimum. Thus the United States, for example, which invented the continental shelf in the Truman Declaration, might have done much better with a much freer system. All this depends in turn on calculations of the exploitation policies of coastal states, and of the international régime respectively, neither of which are easy to forecast. There are of course, in addition, other variables in the positions of developed countries, depending not only on their geographical position, but on their political view-point—how far they are nationalist or internationalist in attitude, for example.

On many of the main points, therefore, national interests differed widely. So, in consequence, have the policies pursued by each state in the discussions. Any régime that is to be agreed must reconcile, so far as possible, these various sharply conflicting interests and attitudes. It is to the way in which this might be done that we must now turn.

CHAPTER 12

A Possible Régime

Types of Régime

We have now considered in this book the main problems—economic, legal, military and environmental—that arise in relation to the sea-bed. We have looked at the negotiations which have taken place on the subject so far. And we have examined the interests of the principal powers and groups affecting their attitudes. We are thus now in a position to consider what type of sea-bed régime might best be able to resolve some of the basic difficulties, best reconcile the various conflicting interests involved, and best serve the needs of the international community as a whole.

This does not mean we shall seek to depict an ideal régime. It would be relatively easy to sketch out the type of model solution which most appeals to the imagination or political predilections, of the writer; or which might be established in an ideal international society, if devised by a group of all-wise, all-seeing and wholly disinterested philosophers. However attractive these might be as a speculative exercise, they would be of no practical value. It is a solution for the real world which concerns us. And any realistic attempt to devise a régime must take account of the facts of that world, the stage of exploitation which has already come about, the course of the negotiations so far, the attitudes of the leading powers, and above all, the basic interests that we have just described. Here, as in most other situations, no ideal solution exists: the solution that is ideal for one set of powers or individuals would be far from ideal for another. We can aim only to define the best possible in the circumstances: the best kind of compromise or balance of interests that could be obtained. It is this which we shall seek to describe.

It may be useful, first, to consider briefly the broad range of alternatives that are open—the main *types* of solution that it would be possible to devise. There are a number of totally different ways of approaching the problem, based on differing conceptions of the main priority, and therefore of the main principle on which the régime should be based. In reaching initial conclusions on these basic alternatives, we may be able to narrow down the choices to be made.

Some of the *theoretically* possible types of régime need not be considered in any detail at all. For though possible, they are in practice either politically inconceivable or intrinsically undesirable—sometimes both. It is possible to conceive of a régime based entirely on free competition, among enterprises or nations, with only the most marginal international control, over such questions as pollution, safety, and freedom of navigation, and, possibly, *voluntary* contributions for international purposes. But this possibility in practice has already been discarded in the course of the negotiations so far, and in the Declaration of Principles; and it would certainly be quite unacceptable to a great majority of governments (though something like it could still come about in practice, by default, if international negotiations were to be long protracted or to reach total deadlock, and might even not be altogether unwelcome to certain states). Again, a theoretically possible régime could come about by a decision to allow unlimited claims by coastal governments to the sea-bed, perhaps based on a wide interpretation of the 1958 Convention, leading to its partition among coastal states. But this would be equally unacceptable to other states and has been repudiated in principle in the Declaration. Both of these could be regarded as extreme laissez-faire solutions. At the other extreme would be a system whereby all exploitation was not merely controlled but directly undertaken by an international authority controlled on the principle of one-nation-one-vote. This solution, too, can be virtually ruled out. Though it could, conceivably, win the support of the majority of nations in terms of numbers, it would certainly not be accepted by some of the largest and most developed powers of the world, and is therefore certainly unlikely finally to be put into effect.

But even when the extremes of this kind have been eliminated, there remain a number of alternative models. First, there is a simple choice to be made between the type of régime that maximizes *revenues* for the international régime within the shortest possible time, and that which maximizes international *participation* in exploitation. If the desire was only to win the greatest possible income for the régime, and so the greatest possible funds for international redistribution, as soon as possible, it is arguable that a system like that advocated by the United States would be the best—that is, a system in which exploration and exploitation were partly under the control of coastal states, and elsewhere almost totally free (because based on licences automatically granted to the first comer) but with royalties paid to the international régime, perhaps on an extensive scale, for all operations undertaken beyond the 200-metre line. This might promote exploitation more rapidly than any other system; and it could create substantial funds for redistribution more quickly. As the main objective of many of the developing countries is precisely to secure large-scale redistribution, it is not necessarily against their financial interests to support such a scheme. Since those companies with the highest technology would be accorded the widest opportunities for exploration and exploitation, such a system could assure the fastest possible identification of resources and the most efficient exploitation of the area.

But though such a system might conceivably maximize the revenues, both for the régime and for many developing countries, at least in the early years, it could do so only at substantial cost. In the first place, it would demand a system of relatively uncontrolled exploitation, which would not necessarily make for the most efficient management and conservation of sea-bed resources. It would be likely to give rise to a frantic race to find and develop the most easily exploited and marketed resources by the companies so qualified, at the expense of efficient long-term development. Secondly, if the trusteeship zone concept was adopted, it would give excessive advantages to coastal states at the expense of the non-coastal. Thirdly, and above all, it would involve making over a large share of the actual exploitation work itself, and of the profits derived from it, to a small number of enterprises from a very small number of developed countries. Whatever the scale of royalties, therefore, and their allocation, the distribution of *profits* would be highly uneven, and would mainly benefit those countries already best off.[1] Even among enterprises, a system of allocation on the basis of applications for sites, though it might do much to encourage exploration, would give huge advantages to those equipped to take part in the earliest years, which would rapidly appropriate the most favourable sites. There is in fact a real dilemma here: to identify the most favourable deposits, exploration should be as free as possible; but this exploration will not take place unless it carries with it also rights of exploitation. It is thus extremely difficult to promote wide exploration without at the same time signing away rights to the most profitable areas. There may therefore be a choice to be made between promoting the most rapid possible exploitation of the resources of the area and achieving an internationally fair system.[2]

If there were an overwhelming shortage of some of the minerals concerned, and the international community's prime interest was to develop new sources of them, the inequities involved in such a system might nonetheless be tolerable. If it could be shown that international economic development generally was likely to be seriously held up without new raw materials from the sea-bed, it might be worth sacrificing a more genuinely international participation in exploitation. Even then, it is arguable that the best course should be rather to encourage the most rapid possible establishment of genuinely international enterprises to share in the work of exploitation. But this is not the present situation. As M. de Seynes, then Under-Secretary-General for Economic and Social affairs, pointed out to the Committee, there is now, if anything, a danger rather of a surplus of copper and manganese, as a result of recent development on land and in the continental shelf, than of a dearth.[3] Sea-bed exploitation, far from being essential to relieve acute shortages of minerals, could cause considerable hardship and economic damage by its effect on the markets of existing producers. And certainly one outcome could be the transfer of production from land sources, often situated in developing countries, to the sea-bed area, operated almost exclusively by enterprises from the most developed countries. As M. de Seynes

[1] However, an African Conference has demanded that concessions off Africa should be awarded only to enterprises 'effectively controlled by African capital and personnel'.

[2] For the simplest way of resolving this dilemma, see p. 226 below.

[3] That is, over the long term: there may of course be short-term shortages, as recently with copper.

pointed out: 'There would be little point in distributing the benefits of such exploitation under an international régime to all developing countries, if they were thereby deprived of some benefits they now enjoyed from the exploitation of their own natural resources, which constituted one of their rare advantages in a fiercely competitive world'.[1]

On these grounds, it is difficult to resist the conclusion that, if a choice of this kind has to be made, it would be better to choose a system in which the benefits were more evenly shared and exploitation more carefully controlled, even if it meant that development took place somewhat more slowly. In other words, if a more fully international system were found to take longer to attain (as is not unlikely) it could still be worth waiting for. Here, as in so many other fields of government, a real choice may exist between economic and social goals. And it is not self-evident that it is economic goals which should always be allowed to prevail.

Let us look at another basic choice between types of régime. There is a choice to be made between those types which afford a relatively large role, and a large area, to the coastal state but a relatively strong type of international régime beyond; and those that favour a narrow margin for the coastal state but a weaker system in the international area (providing therefore a considerable role for national states or private enterprises within that area).

These are not *logically* alternatives. It would be perfectly possible to have a system with a narrow margin and a strong régime; or a wide margin and a weak régime. But in practice the two described probably do represent the range of political possibilities. This is partly because nearly three-quarters of the nations of the world are themselves coastal states, which would mainly benefit from a fairly wide range of coastal jurisdiction over resources; and these include some large and powerful states with a strong chance of maximizing their interest through retaining control of a fairly wide continental shelf. These powerful states may be ready to renounce the chance to control their entire continental margins only if they are assured that they will be provided with reasonable opportunities within the international zone beyond—which is only likely to happen under a relatively weak régime (in the sense used here). Conversely, such countries

[1] A/AC138/SC2/SR35. This argument of course applies only to the producing countries; other developing countries might gain from reduced prices for sea-bed minerals, and more widely available supplies.

would be willing to accord strong powers to the international authority in the international area only if they were assured of a relatively wide area still under their own national control. Against this, there are a number of developing countries, and especially the land-locked and shelf-locked, which have an interest in a system of both narrow margins and a strong régime beyond. They, therefore, will probably only be willing to concede wide margins in return for a strong system outside, or vice versa. At root, as we saw, the basic difference of interest is between the coastal and non-coastal states. It seems almost certain that the final outcome will be some kind of compromise or balance of interests between them.

In fact, because of the complexities of the various and diverging interests involved, it seems likely that there will be a double compromise; not merely wide–strong against narrow–weak, but a compromise which comes fairly near the middle of the range of possibilities in both respects. Thus the exploitation boundary will be wider than the minimum possibility of 200 metres in depth, but it will, one hopes, not be as wide as the maximum, under the U.S. trusteeship proposals, of about 3,000 metres in depth (the edge of the continental margin). It now seems most probable that it could be somewhere around the limit so strongly proposed by the Latin Americans (provided coastal state rights are carefully circumscribed). Certainly it could not now conceivably be so narrow as the 30 or 40 miles that some of the land-locked countries favour. Similarly, so far as the *strength* of the régime is concerned, it will not be such a nationalist system as the Communist states would like. But it will also not be so internationalist as the Latin-American proposals suggest, with direct exploitation by the international enterprise. Here, it seems likely that the compromise may be based not on a mid-point but on a combination of proposals; that is, there could be provision for participation by private enterprises, by national governments, by regional groups, as well as by an international enterprise in the international area. And whereas national governments may have a role in vetting and controlling production, there will also be a considerable work of supervision for the 'international machinery'. We will consider later the exact form which these compromises might take.

The next range of possibilities concerns the authority itself, and the system of decision-making within it. Here, if the extreme possibilities are again eliminated, the effective choice is between an authority that is highly 'democratic' (one nation, one vote) but whose powers are limited, and one that is non-democratic (with weighted voting) but having far stronger powers. This is once again because a considerable number of countries, including some of the most powerful—especially the United States and the Soviet Union—will be willing to accord significant powers to the authority only provided that voting within it is weighted. Conversely, if the authority is established on a 'democratic', or even a moderately 'democratic' system, they will not be prepared to concede it worth-while powers. And they would certainly never enter into a treaty in which 'democracy' and strong powers were combined. It is even possible that the two super-powers will do all they can to ensure that the authority has neither of these attributes: that it both operates according to a system of weighted or balanced voting, and that its powers of discretion are limited, or at least rigidly defined.

The effective range of alternatives has in fact been indicated by the proposals already put forward. At one extreme there is the Soviet demand for 'consensus', with the agreement of every member of the proposed Council needed for all decisions. This is run fairly close by the U.S. suggestion that a two-thirds majority among the six most industrially advanced states, as well as within the Council as a whole, would be required. At the other extreme is the Latin-American proposal under which the Assembly, representing every member, would be able to take decisions on any subject by a two-thirds majority. Here again a compromise is likely. There are a number of alternative ways in which an element of democracy could be combined with an element of consensus, and some of these will be discussed in a moment. At present, it is sufficient to establish that, once more, the final outcome is likely to come somewhere along the middle of each range. The authority will take decisions according to a system that is not wholly undemocratic or veto-bound, yet gives no total domination either to developing or developed states. And it will have powers that are relatively wide, yet at the same time are very carefully defined.

This leads to another distinction between possible types of régime which needs to be considered: between the 'ocean space' approach, favoured by Malta and some other countries—that is, a régime having powers in many fields throughout the oceans, and not just over sea-bed resources—and a system strictly confined to sea-bed exploitation. Here the decision will not necessarily be reached on the basis of national interest. So far, the developing nations have perhaps been marginally more in favour of the ocean space concept; but this is really only a reflection of their general desire for a strong system. If an equally strong system were proposed, but with authority divided among a series of separate agencies, each responsible for different aspects, this might be equally attractive to them. The choice is here really one of judgement rather than interest. And it is thus one of the few issues that may finally be decided on the basis of the collective wisdom of the international community, exercised in a relatively impartial and disinterested way, rather than by the self-interested calculation of national interests and a long and bitter process of tough bargaining to find a compromise between them.

There is no doubt that the conception of a unified system of regulation for the whole of ocean space, under all its aspects, is a highly attractive one. It recognizes the interrelated character of all activities in the area, the complementary interests which nations have within it, and the need for an over-all view of the way in which the area should be governed. It has won high praise among conservationists and in the British House of Lords. Yet this is nonetheless one of those concepts which, the more closely they are examined, the less convincing they become. The fact that activities are closely interrelated is of course a strong reason for ensuring careful *co-ordination* of all decisions and actions affecting the area. But it is not necessarily a reason for having one single super-government, responsible for every aspect of human activity there and for every decision reached throughout that vast territory. Mineral exploitation, agriculture, conservation and arms control are sometimes interrelated on land. But nobody suggests that for this reason there should be one single national ministry, or a single international agency, responsible for all of them. Nor does anybody suppose that the UN agencies concerned with agriculture, education and labour should be merged within a single entity simply because their activities very often interact and overlap (and often give rise to bitter disputes among them). The answer in all cases is a system of co-ordination rather than unified control.

The paradox of the ocean space approach is that it *under-estimates* rather than overestimates the complexity, variety, and scope of ocean activities and interests. Given the scale on which these are now developing, to have one single authority responsible for all would be rather like suggesting that there should be only one single ministry responsible for all human activities within a state; or only one UN specialized agency responsible for all international problems and activities on land. Any agency that attempted to supervise the entire range of marine activities, including the monitoring of pollution, managing the oceans' many different stocks of fish, organizing arms-control activities, overseeing scientific research, besides supervising, regulating and directing the many different kinds of exploitation of underwater resources, not to speak of administering and controlling the human habitats which may shortly appear within the oceans, would represent a vast monstrosity of bureaucratic organization, which could make existing international organizations, themselves often cumbrous and over-extended, appear like the most streamlined of parish councils.[1]

What rather seems to be required is the establishment of separate and specialized administrations responsible for different aspects of ocean activity, but with the maximum possible interrelation between them where activities converge, together with an effective system of over-all co-ordination. The latter could be done in a general way by ECOSOC and the General Assembly. But yet better would be an Oceans Council, more specialized in responsibilities and skills, of governments and the agencies concerned, which would exercise an over-all view of activities within ocean space, and ensure that effective co-ordination among them was achieved. This would provide the essential advantages of the ocean-space approach without the heavy cost in bureaucratic top-heaviness.

There is one final basic choice between types of system to be made. This too is a choice which could determine many other features of the régime, and is therefore of fundamental importance. It is between a system in which licensing is to states in the first place, and one in which it is direct to the operating enterprises.

It is perhaps understandable that during the discussion of this issue it has been so widely assumed that licences must be given

[1] The Maltese draft treaty proposes that IMCO, the IOC, and the FAO's Fisheries Committee should be 'consolidated' within its proposed institutions. It also suggests that the institutions should take over the responsibility for administering certain categories of islands, including inhabited islands, with the consent of their peoples: quite a sizable role.

in the first place to governments; and that governments should play a dominant role in administering the area—in vetting enterprises, in undertaking inspection, in controlling pollution, and in receiving revenues. Discussion has taken place in a world that is still dominated by the nation-state. Quite apart from the natural bias among the representatives of such states in favour of their own capacities and claims, it was a natural presumption that, in any activities on the vast scale of sea-bed exploitation, governments must have a large role to play. They play a large role in every other form of human activity. International organizations have never assumed such powers as would here be required. There was reasonable doubt about the capacity of any such organization to undertake that work. There was a desire that governments should have a share in the revenues. It was not altogether surprising that, among those taking part, it should be so widely assumed that national governments would have a large part to play in administering and controlling this vast and valuable region.

Yet in making that assumption, many of the difficult problems which would anyway arise were automatically enormously compounded. A number of the most contentious and difficult aspects of the problem are made far more difficult if the intervention of governments is assumed; some of them scarcely arise at all if no such intervention takes place. Such questions as the division of the area, the criteria on which allocation is made, the limit to the area held by each state, the balance of authority between national governments and the international authority in administering the area, the initiative, all of these (which we have already seen have been the subject of bitter dispute) occur only on the assumption that national governments are given responsibility for specific parts of the sea-bed. By removing that assumption, a whole range of difficult problems are to a large extent eliminated.

Part of the reason is simply that national rivalries are the most intense and profound within the modern world. To superimpose them on the already difficult problems of the sea-bed inevitably is to compound the difficulties. For example, the allocation of blocks for exploitation (and deciding the principles that shall govern it) immediately becomes a far more acrimonious and politically dangerous issue if it is an allocation to nations, each determined not to be outdone by other nations, than if it is allocation direct to companies. The question of royalties and the division of revenues becomes far more acute if it concerns not merely a division between the international régime and the profits of enterprises, but a division among nations as well. The problem of the delimitation of exact boundaries of blocks immediately becomes far worse when it is the boundaries of national territories that are introduced. The question of the initiative becomes insoluble. Conversely, a decision to award licences direct to enterprises (public or private) has the opposite effect. Régime, enterprises and nations, all then have a common interest in the maximum possible exploitation and the maximum possible turnover. Though their interest is not identical, the problems, as we shall see, are far less.

There is a second series of reasons why a direct relationship between the sea-bed authority and the enterprise has every advantage over one providing an intermediary position for governments. The latter system creates a vast higgledy-piggledy patchwork of separate jurisdictions and different systems of control, varying from block to block, all across the sea-bed. This would not only cause serious practical difficulties in establishing effective international regulation, it would also mean that in some cases standards would be quite inadequate, either because a state is not capable of administering the complex regulations required, or because competition among governments to secure the most rapid possible exploitation induces slipshod or lax regulations. Above all, there would be a duplication of national and international regulations. This might mean that neither would be adequate. Measures against pollution, for example, if they are to be of any value, must be imposed on a uniform basis and apply uniform standards. To suggest that national governments might wish to impose higher standards in their own areas if they wished, as some have done, means either that the international standards would be too low, or the national standards too high (that is, too protective). This applies equally to standards concerning safety, technical capacity and the financial backing of enterprises. Duplication here is an absurd extravagance. Assuming (as is almost universally assumed) that there must be comprehensive international regulation on these subjects, there can be no advantage, and manifest disadvantage, in providing for national governments to impose their own standards and controls in addition. All this applies above all to the process of inspection. It is everywhere accepted, by developing and developed alike (except possibly by the Communists), that the international authority will require to have its own inspection service to enforce the standards it lays down. The creation of an additional system of inspection by governments, far from being a 'saving' of expenditure, as some delegations have suggested, would on the contrary be an astonishing and lunatic waste of resources, especially given the very substantial costs involved in even a single inspection of a particular site in sea-bed areas.

Thirdly, the carving up of innumerable tiny pieces of the sea-bed among individual governments has more general disadvantages. It would represent a perpetuation of the nation-state system of the most undesirable and unnecessary kind. It would establish what is in effect a new form of colonialism within man's final frontier on earth. That frontier could provide the test-bed for a wholly new type of international system providing for co-operation rather than competition among states, and a new means of redistributing wealth among them. If there is a single valuable asset which the oceans have inherited from their largely neglected past, it is surely the international character that has always been accorded to them. 'High seas' have been always synonymous with 'international seas': seas that are shared equally by all and can be reserved by none. This non-national character of the surface of the waters has until now been extended equally to the sea-bed. And yet the very countries which have always been, rightly, the most strenuous and vociferous in asserting the international character of the seas' surface are now the busiest in trying to parcel up its bed into innumerable pocket-handkerchiefs of national sovereignty all around the world.

There are also obvious practical defects. Such a system would radically reduce the revenue available to the international system itself. The net profits from the sea-bed, over and above the operating cost and a moderate return on capital for the enterprises, instead of accruing direct to the international régime, would be shared between that régime and large numbers of national governments. The funds available for redistribution to poorer countries would thus be reduced. There would be conflicts between governments and the régime concerning the proportion of revenues to go to each. There would be continuing disputes not merely among governments competing for areas, but between governments and the authority concerning the nature and range of the authority's powers, or concerning the application of particular regulations. There would be scope for bitter conflicts between the governments administering particular areas, the companies to which they offered licences, and the governments supporting their companies. There would be complex and confused disputes between two or three governments with rights in adjoining areas, the authority, the companies undertaking the operations, and the governments of the countries from which the companies come. In a word, the difficulties that must anyway arise concerning rights in the area are enormously compounded and sharpened by the interposition of national governments, and so of national rivalries, with all the special intensities with which they are associated.

However, the final, and by far the most important defect of such a system is different again. We have already seen that the actual work of exploitation will anyway be undertaken by large commercial enterprises which alone have the necessary technology. The type of system established will make a vital difference to the conditions under which such countries negotiate for contracts and royalties, and the terms that they can obtain. If allocation is to national governments in the first place, a relatively small number of qualified companies will be able to offer their services to 140 or so different national governments. The governments, very few of them having companies of their own able to operate, will thus be in the position of *competing* to obtain the services of the companies, and so the technology which they alone could offer, for their own areas. The bargaining power of the companies would in such circumstances be extremely strong, and that of the governments very weak. Some governments might at first hold out against offering a concession to a foreign company on the terms it demanded, hoping perhaps to wait until it had developed enough expertise among its own nationals, or until some new regional enterprise came into being to undertake exploitation. But when it saw its neighbours acquiring quick revenues with the help of a foreign company, it would probably soon sink its pride and offer a contract to such a company on whatever terms it could. These would be likely to be terms highly advantageous to the companies. Among the small number of companies concerned, competition might well be less than perfect. Some might prefer to deal only with their own governments, or the governments of developed states, or governments that they approved. Other governments might thus be unable to take advantage of the blocks which they had acquired except on terms very unfavourable to themselves and very profitable to the companies.

If, on the other hand, licensing were direct by the authority to the companies, this bargaining situation would be reversed. The qualified companies could then gain access to the resources only with the consent of a single authority, which would be responsible for all negotiations. Instead of governments competing with one another for companies, companies would now be competing with one another to obtain sites from the authority. The authority would thus be able to award licences, both for exploration and for exploitation, on terms which were far more favourable to the régime than under the alternative system. It could, if it wished, introduce a system of competitive bidding for favourable sites, as national governments very often do within their shelves—a system which would be quite impossible if there were allocation to governments, as in that case the richest governments would acquire the best sites (this is another of the problems removed by direct licensing). Clearly, the terms offered to companies would still need to be adequate to assure them of some profit after the expensive investment costs had been covered. But this could sometimes take the form of a share in the over-all profits, or even a fixed fee or commission. In either case, the windfall or 'rent' element in the profit, resulting from the scarcity of technology in relation to newly discovered resources, would fall to the régime and not to the individual enterprise. All the bitter conflicts which would take place among states concerning the allocation of blocks and the principles on which it should be based if allocation were to nations, would here be removed altogether. Finally, under this system all royalties would go to the régime, to be distributed on a progressive basis, instead of having to be shared with governments all enjoying equal rights.

In almost every respect, therefore, a system of direct licensing to enterprises (whether private, national or regional) would be superior to one in which national governments intervened in the process. It is thus curious that among the three main possibilities this is the one that has been least considered in the international negotiations so far. The two main alternatives—licensing to governments or direct exploitation by the international enterprise itself, each of which have manifest disadvantages—have been far more discussed. Those need not necessarily be altogether excluded by the terms of the treaty. But there is no doubt which system should be the one primarily envisaged for the régime to be established.

The System of Exploitation

This consideration of the relative advantages of the *types* of system available has somewhat narrowed down the available choices to be made concerning the régime.

It has led us to certain general conclusions: that the maximization of revenue for the régime in the early years need not necessarily be the decisive objective; that the system chosen must afford to the coastal state a reasonable but not excessive breadth of sea-bed under its own control, and to the international régime reasonable powers of control over the remaining area; that the authority should reach decisions on a basis that is not 'one nation, one vote' but not inhibited by veto power, and should have powers which are reasonably broad but carefully defined; its powers should be confined to the exploitation of sea-bed resources, but it should co-operate closely with other agencies having other responsibilities within ocean space; and, finally, above all, licensing should be direct from the international authority to the operating enterprises.

Let us now look in a little more detail at the system of exploitation required.

The first point to note is that the system need not be uniform. The resources of the sea-bed are of a number of quite different kinds. It includes the three main categories often referred to in UN debates and in some of the draft treaties: fluid sea-bed resources, such as oil, gas, sulphur, and geothermal energy; nodules on the surface of the sea-bed; and hard minerals beneath the sea-bed. But it includes, in addition, others that are rarely mentioned: the phosphorite and other resources of the surface of the sea-bed; hot brines in a few isolated areas; the living resources of the sea-bed (that is, the so-called sedentary fishing); and gravel, copper and tin, at present dredged near the coast, which may eventually be exploited in deep waters. Much of the UN discussion has seemed to be based on the assumption that, apart from slight distinctions relating to the size of blocks allocated for different resources, a generally uniform régime must be applied. There is in fact no necessary reason why this should be so. The conditions relating to dredging, for example, requiring a considerable degree of mobility over a wide area but a relatively brief period of active operations, might demand a totally different system from those relating to oil and gas, which require fixed installations for a considerable period of time. Thus, in what follows we shall have to consider at least the possibility of wide variations in the systems applied to each different raw material.

It is perhaps logical to consider the system in the order of operations. The first problem thus relates to exploration. The major difficulty here is that exploration for commercial purposes is difficult, if not impossible to distinguish from that taking place for other purposes—for example, scientific research. Purely scientific exploration could always have a commercial application. Many other activities, including navigation in submersibles, underwater photography, magnetic and gravity surveys, and others, can be used for commercial exploration without special permit as they are anyway allowed under the normal law of the sea. But the knowledge acquired in this way could bring important benefits to those who had undertaken them when applications were made for exploitation rights in particular areas. In any case, to attempt to restrict and control commercial exploration of this kind could prevent knowledge of the most valuable and easily accessible resources being obtained, and hold back economic progress and the régime's revenues.

Some of these problems are already removed once a decision has been made, as here suggested, to award licences direct to companies and not to governments. The special advantages to be acquired through intensive commercial exploration become far less serious. Freedom to explore would no longer give special advantages to particular *nations*. And companies could all equally benefit. Under these conditions the greater the special knowledge acquired by each enterprise and the more rapid the subsequent exploitation, the better for the international community as a whole. The fact that exploration could be relatively freely and easily undertaken becomes, not a liability, but an asset. And the need to make unenforceable distinctions between exploration of different types would disappear.

This suggests the need for a licensing system similar to that used by many national governments. Under this (for example, in Canada) enterprises can obtain, relatively easily and for only a small payment, an *exploration licence* giving the right to undertake perhaps any activity other than deep drilling, and over a very wide area, possibly the entire ocean bed.[1] This would not in itself accord the right of exploitation. For this, application would have to be made for a separate *exploitation licence*, for a particular mineral in a particular area; and a far more substantial payment would then be required. If desired, there could be some intermediate category of right, an *evaluation permit* which would give the right to undertake more intensive evaluation of a given deposit in a particular area than could be undertaken on the exploration licence alone. The exploration licence would be for a relatively short period—say, for two or three years—but could easily be renewed; and the evaluation licence perhaps for five years. The exploitation licence would be for a much longer period, as security is here essential; possibly fifteen years, but subject to renewal, when new terms could be negotiated.

There are nonetheless still certain difficulties. An exploration permit need not carry with it an *automatic* right to exploitation of any finds made; the Canadian and U.S. systems, for example, do not give this right. But it must at least carry a strong probability that such rights will in practice be obtainable once finds have been made. If this were not the case, no company would regard it as worth undertaking the enormous costs involved in exploration work. This means that, in effect, in awarding an exploration permit, the international authority is also at the same time making a half-promise of the right to exploit in part of the area explored. In so doing, it is losing a part of its discretionary powers, and of its control over the area. If the main aim were simply to maximize production and revenues, no great difficulty would result. Every enterprise would be encouraged to undertake the maximum possible exploration, and would be awarded production rights on the basis of its finds. It would then be left to the forces of competition to show which could produce most at the cheapest prices and at the greatest profit to the authority.

But we have already seen that the maximization of production

[1] *Exploration* is the process of prospecting by reconnaissance surveys, geographical, geophysical, hydrological and topographical, sometimes including scout drilling and sampling. Rights for this are usually given on a non-exclusive basis. *Evaluation* includes sampling, drilling, testing and feasibility studies. Here exclusive rights are usually given, and they normally carry with them the right to obtain an exploitation licence at a later stage if so desired. *Exploitation* includes concentration, processing, refining, separating, smelting and treating, as well as mining or drilling. Sometimes evaluation and exploitation licences are combined in a single exclusive permit, as in the United

is not the only aim. Spreading opportunities to work the sea-bed is another. If the authority wishes to diversify the enterprises engaged—to protect land production, to conserve resources, or for any other reason—it may wish to restrict production. In this case, it could need to refuse exploitation rights, at least to a particular company. This in turn should lead it to refuse exploring rights as well (as these imply a chance to exploit). It could limit the exploration licence to a relatively small area, so that the chances of a find were reduced—although this loses a large part of the advantage of the system, which is to promote exploration over as wide an area as possible. Alternatively, the authority could limit the number of operators to which it gave even exploration permits; but this could raise difficult problems concerning the choice of enterprises. And since the enterprises would almost certainly be more or less closely associated with particular nations or groups of nations, acute political difficulties could arise over any such decision. The best course might be to give *restricted* exploitation rights on the basis of finds. The authority would need to relate exploration permits to ultimate production needs. And it would need to ensure that equal opportunities of profitable exploitation were accorded in this way to different enterprises from different regions, whether private, national or regional.

This leads to the other major difficulty. This concerns the criteria according to which particular enterprises would be selected to undertake exploration and exploitation. If the main object were to maximize the revenues of the system, then presumably selection would be on the basis of technical capacities alone. Licences would be granted to the enterprises which seemed likely to bring about exploitation most rapidly and efficiently. A system of auctioning rights might be the best way to achieve this, and would also maximize revenues. But this would mean in effect that most licences would go to the largest and richest corporations, which would mean primarily North American or West European. And it may be doubted how far this would be politically acceptable to many of the members of the system.

Under a system of direct licensing, the award of contracts would be less controversial than if licences were granted to governments, but it would still have some significance. However strong

---

States, Britain and other countries. Sometimes all three are combined in a single exclusive licence, as in New Zealand, Argentina and other countries.

the international authority's bargaining power, and however low the return on capital it allowed to the operating countries, it would remain the case that the nations to which the operating companies belonged would acquire some benefits over and above their normal shares of the revenues. In this situation the authority would, no doubt, like many national governments issuing contracts, seek to exercise its discretion in ensuring a reasonable balance between enterprises of different regions and types. Alternatively, it could seek to ensure that every operating company included some participation from different countries and continents. Finally, the most extreme in fairness (but also the most economically irrational way) would be to assign licences by lot (as is sometimes done by governments). Whatever system was used, a company, in order to qualify, would probably have to undertake to employ a certain proportion of nationals of developing countries, or to undertake some training programme for them.

Decisions of this kind would involve a certain element of arbitrary choice and discretion, and the abandonment of the objective criteria which many countries have demanded. Such choices would be semi-political. Considerable conflict and controversy could be aroused. Though the award of contracts direct to operators rather than to governments would reduce political conflicts, therefore, it would be a mistake to think that they would be eliminated altogether: that a wholly impartial system for awarding contracts is possible. The best that could be hoped is that the system applied would be neither wholly economic nor wholly political; that there would be stringent tests of the economic and financial capacity of every enterprise awarded a contract; but that there would also be an attempt to achieve a balance between countries and continents of origin. This is not very different from what many national governments do in awarding sites in their continental shelves.

Under direct contracting to enterprises, the question of 'initiative'—who decides which areas should be exploited and when—also becomes of less importance. The authority has an equal interest with the operators in ensuring that the most favourable deposits are worked as quickly as possible, and normally it would encourage applications for any area on the basis of explorations undertaken. Only where two enterprises had surveyed precisely the same area are conflicting applications likely to be made. The authority would, perhaps once a year, consider all the applications made by different operators for rights in different areas. On the basis of a decision in principle concerning the amount of new exploitation required for each resource, and the balance between enterprises and their nationalities, it would then announce its decisions on each application. This could represent a proportion, say 30 per cent of the total rights applied for, relating to any particular mineral. Unless normally a part at least of the rights asked for were granted, there would be no real incentive for exploration to take place. The authority would not only seek a balance among operators, taking into account also their assessed operating efficiency and financial backing; it would also probably apply a maximum to the area which could be held by any one enterprise, whether private or governmental. Thus the initiative would in effect be shared between the operator and the authority.

This would be the régime for oil and gas. In the case of the nodules, a different system might prove preferable. Here the technique of exploration, though still beyond the capabilities of most nations and enterprises, is less complex and expensive than that needed for discovering sub-surface deposits. The resources are accessible and easily visible on the surface of the sea-bed. All that is required to discover favourable sites is wide-ranging under-sea photography, combined with a certain amount of sample dredging and subsequent analysis. The main difficulty and the expense is thus not in finding the deposits but in building the enormously powerful and complex equipment to dredge from great depths, and to extract the different minerals after they have been recovered. Another difference is that for such resources there is less importance in the acquisition of one site rather than another. Since oil and gas occur irregularly and only in certain situations, it is vital for the exploiting operator to acquire one particular site rather than another, and to obtain rights there over a prolonged period. In the case of nodules neither is important. Once the operator has built his equipment, one area of the sea-bed may well be almost as favourable as another. Nor does he require a long period of tenure in any particular place to be able to operate effectively. Indeed, his interests (and that of the régime as a whole) may be best served by allowing him the maximum freedom to travel around the sea-bed relatively freely, so that once he has exhausted one area, or made no finds, he can move on to another.

Here, licences for exploitation do not need to be related to any particular area. The operator would merely be required to make a large payment (perhaps based on the capacity of his equipment) related to the amount which he could be expected to exploit in any part of the ocean during a particular period. He would require some assurance that he would not be disturbed or challenged in his rights to any particular deposit on which he was temporarily working. But this could be provided by a general rule that no operator could go within a certain distance—say, a mile—of another operator's activities, of which the latter had given previous notice.

Sub-surface hard minerals are unlikely to be worked in the international area for many years.[1] But they too might eventually require a separate type of régime. This might need to grant the operator tenure which was longer and even more secure than that for oil and gas. Mineral resources on the surface, such as tin, iron and iron sands, could be dredged and relatively easily explored, like nodules, but in this case would probably need a system of allocation in blocks, perhaps relatively small but for a considerable period; or perhaps *ad hoc* licensing in response to a particular application. Sedentary fishing resources should perhaps be supervised by a fisheries commission, rather than the sea-bed authority, and in any case would probably be of interest only to a relatively small group of fishermen of geographically neighbouring countries; here exclusive rights should not be given, subject only to the provisions of the 1958 Convention on the High Seas concerning long-standing prescriptive rights. Gravel and phosphorite deposits are unlikely to be exploited in the international area in the near future; but could also require a separate system of their own eventually.

For all these resources, general undertakings would need to be given by operators concerning their operating practices: to avoid causing pollution (this would be subject to heavy damages, for which an insurance system might be introduced); for avoidance of unreasonable interference with other uses, including freedom of navigation; for the protection of living resources and marine life; for adequate safety regulations; for the effective marking of operating areas and notification to other users; for the prevention of undue waste of resources or damage to the marine environment; for the obligation to give rescue and other assistance to

[1] At present, off-shore mining of these minerals is almost all at depths below 60 metres.

other operators in case of emergency; and for liability to civil damages. Operators would be required to submit work plans to the authority, for exploration as well as exploitation, and to submit regular reports once exploitation had begun. They would need to supply geophysical data and other information acquired on a confidential basis to the authority, and they would have to undertake to submit to inspection by the authority.

The primary responsibility in all these respects, financial and otherwise, would lie with the enterprise itself. But it would be for consideration whether the national government of the state in which the enterprise was incorporated, or, where appropriate, a group of such states, should take on themselves an ultimate responsibility to accept liability in case an enterprise, because of bankruptcy or other reasons, totally failed to meet its liabilities or violated the regulations. The main sanction available to the authority to secure compliance would be the threat of forfeiture by the enterprises concerned of all existing or future rights within the area.

A number of different types of payment would have to be made. A relatively modest fee could be charged for an exploration licence in the first place. Where an exploitation licence was obtained, different types of payment might be demanded. First, a substantial lump sum could be paid over, perhaps as a deposit: this would act as some incentive to the enterprise to provide sufficient investment in the area and secure a return as early as possible. Then there would be some kind of annual payment, rising progressively until production began (both these are features of many national systems). Once production began, royalties would be paid at a general rate laid down, or a rate negotiated, calculated on the estimated value or volume of production. Finally, there could also be a payment based on profits which would correspond to the tax payment that would be made in any national area (this assumes that there would be no taxation to be paid to governments).[1]

Under the system here described, the level of royalties and other payments would be set by purely competitive forces. Because enterprises would in effect be competing for licences, they would

[1] Governments might be required to enter into undertakings concerning the tax treatment which they would give to enterprises operating within the sea-bed area: pledging themselves not to subsidise such enterprises, and promising to give normal customs treatment to production from the area.

be willing to accept them for whatever they regarded as the minimum rate of return on capital acceptable. So long as there remained any enterprise, private or public, still willing to operate and able to operate efficiently, the authority could continue to maximize its own revenues. Moreover, it could, if it so wished, seek to resort to alternative arrangements, such as engaging companies to exploit under contract for a fixed fee or percentage, or to enter into joint ventures with the authority itself. Obviously some profit would need to be allowed to each company and the level of royalties would have to be set at a level that did not deter exploration and exploitation altogether.

As in some national systems, the deposit might be recoverable upon evidence that a certain level of investment had been undertaken. Certainly, some assurance concerning levels of investment and of production might be required. But to a large extent a high initial payment already operates to provide a guarantee of this kind.[1] There would also need to be undertakings concerning levels of production, once exploitation had begun. In the case of failure to achieve levels of investment or of production, the enterprise would risk forfeiting its rights in the area (unless there were special mitigating circumstances). Such relinquished areas could be relicensed or auctioned. It should also be possible for the operating enterprise to relinquish, or to transfer, areas in return for payment of a certain sum. Finally, there could be provision for the automatic relinquishment of a proportion, say 7/8ths, of an area held, and not worked, after a certain period, so that it could be relicensed or auctioned to other enterprises.

Though licences would be subject to re-negotiation after a fixed period, they would clearly need to provide reasonable stability for the enterprises for the period of its duration. In a more long-term sense, too, the system would have to appear stable, for companies would probably be unwilling to undertake the large investment required if they felt that the rules might be changed

[1] In the Canadian continental shelf, companies pay a fee for each permit at the time it is issued, and also deposit cash, bonds or a promissory note equivalent to the full amount of the work requirements for the first period of the permit, as a guarantee that the work will be carried out. Similar guarantee deposits are made also for each succeeding period of work. The deposits are returned upon receipt of satisfactory evidence that the appropriate work has been performed. The work requirements increase progressively during the period of the permit so as to reflect the progressive increase in expenditure necessary to evaluate an area effectively.

within the not very distant future. Subject to that proviso, it would be open to the authority to review from time to time the way in which the arrangements were working, in relation to the management and conservation of resources for example, and to make any necessary adjustments to the system.

It is because it might have this power to revise or change the rules of exploitation, that special importance attaches to the structure of the authority and the way in which it reaches its decisions.

The Institutions of the Authority

Having built up a general picture of the system of exploitation that might be used, we are now in a position to consider the kind of institutions that would be required to implement it.

It is widely accepted that the central institution would be an International Sea-bed Agency, responsible for the general regulation of the area. Much of the day-to-day work would be performed by a permanent secretariat. This would undertake the initial examination of all applications for licences; study the technical and financial capabilities of the various enterprises applying; and on this basis make recommendations to the agency. It would organize the inspection of operations to ensure that operating standards were being fulfilled. It would collect the different fees, rentals and royalties, organize the agreed distribution of the proceeds, and generally manage the finances of the agency. It would supervise the assumption, relinquishment and transfer of blocks at the proper times. It would issue regular reports to the political bodies, and service their meetings. It would organize the training and transfer of technology undertaken by the agency. And it might organize some independent research, exploration and technical studies, to the extent necessary to enable it to perform its duties effectively. It would be under the direction of an executive head, comparable to the Secretary-General of one of the specialized agencies—though in this case perhaps the title managing director, as used in the IMF, or director-general, as in the ILO, might be more appropriate. This managing director would exercise personal managerial authority in what would be, at least in part, a functioning commercial organization.

All this is relatively uncontroversial. The real problem concerns the system of political control. It is generally accepted that there would be an Assembly and a Council, as in the specialized agencies. The Assembly would include representatives of every member state, would consider matters of general policy, and would meet relatively infrequently—perhaps once every two years. The Council would meet more often: several times a year, or perhaps whenever needed, like the IMF's executive board. Like the Councils of the specialized agencies, it would have perhaps 20 or 30 members (the unhappy tendency to inflation among these councils recently should be resisted) and would undertake most of the day-to-day running of the organization. These political bodies would of course take the major decisions: for example, concerning the allocation of sites, the scale of royalties and the distribution of revenues.

The main problem concerns the voting system to be used in these two bodies, and the relative powers to be wielded by each.

The developing countries, as we saw, have called for a system of one nation, one vote, whereas the United States, the Soviet Union and other developed countries have as strenuously resisted this. Even if it was politically conceivable that the latter would give way on the issue (and it is not), the theory that 'one nation, one vote' is a more 'democratic' or more 'representative' method than others is open to serious question. What it would really mean is that the 100,000 people of the Maldive Islands would have an exactly equal say in the agency's decisions with the 200 million of the United States and the Soviet Union, or the 800 million of China; it would mean that the 200 million people of Africa would have 40 times as much say as the 200 million of the United States; that the 70 smallest nations in the world would wield a majority in the organization, though together they have only a sixth of the population of the two largest members. And so on. This may preserve the principle (if there is one) of 'sovereign equality' but certainly not of democracy. The concept of sovereign equality can make some sense in bodies such as the UN General Assembly, which are designed to present the diversity of world opinion on any subject but take no binding decisions: still less executive decisions. It is totally unfitted for executive bodies, and least of all for a body which can exercise such wide-ranging and crucially important powers as the proposed sea-bed agency.

What is required is to find a way of securing decisions that are broadly representative of majority opinion but also give adequate weight to the opinions of major powers or groups. Within the UN specialized agencies, a variety of procedures have been developed to achieve this.

The first is through a system of weighted *voting* to give greater influence to larger or more economically powerful states. This is the system used, for example, in the IMF and the World Bank, where votes have been based on complicated formulae related to national income, shares in world trade and other factors. These systems are often heavily criticized. But in fact it is arguable that they provide a balance of voting strength far more closely related not only to power but to population than the system of 'one nation, one vote'. The weighting need not, of course, as these, be based on economic strength alone. Certainly in the sea-bed context, it might be better to have weighting, on a modest scale, according to population: for example, giving five votes to the United States, the Soviet Union, China, and perhaps other very large countries (say over 100 million population), 4 to nations of 20 million or more, 3 to those of 5 to 20 million, 2 for those of 1 or more, and 1 to the very small. This would be a healthy precedent for other bodies. However, hostility to any system of weighted voting is so great within the international community that it is most unlikely that any system of this sort will in practice be adopted. Nor could it in itself always make sure that every decision was based on a widespread consensus: it might simply mean that the views of many small nations were consistently ignored.

Another system used in the specialized agencies for this purpose is weighted *representation*. In the International Labour Organization, for example, a proportion of seats in the governing body are reserved for the most industrially developed countries; in the Intergovernmental Marine Consultative Organization, for the chief shipping and ship-using countries (both mainly highly developed); in the International Civil Aviation Organization, for the nations of chief importance in aviation, and so on. In a sense the U.S. proposal calling for permanent representation of the six most industrially advanced countries in the council of the sea-bed agency is a variation on this theme; though (with the proposal for a group veto) a somewhat extreme one. But this system involves a form of privilege which is particularly conspicuous, and so particularly unacceptable to most developing countries. These are certainly unlikely ever to accept the proposal in the form put forward by the United States. But unless it occurs in almost as extreme a form as this, it would not serve its main purpose:

ensuring that all decision are based on a fairly wide consensus, and are not totally unacceptable to major groups. There would certainly be great difficulty in deciding which countries were 'important' enough for special representation.

A third system used in the UN family for this purpose is to accord *veto power* to particular nations, to ensure that their interests are not totally overridden: as in the Security Council, or in the League's rule of unanimity. This too would be theoretically possible for the sea-bed régime. The Soviet proposal for decision by 'agreement' would provide a universal veto. This could be amended to provide a veto only for the largest powers of all, as in the Security Council. But this again is an extreme solution, which could have the effect of reducing the agency to total impotence because of the recalcitrance of a single state. Somewhat more viable would be a system of group vetoes, under which a majority of each geographical group was required for all decisions of a certain kind. Under this system no nation alone could prevent a decision, but a variety of interests would need to be reconciled for any to be reached. But even this could be a crippling restriction on effective action (especially in view of the fact that one of the groups would be the Communist block, not notably positive in its attitudes on this issue).

But there are further possibilities which are perhaps more likely to provide the solution required. One would be a system similar to that which has been established in the UN Conference on Trade and Development. There the original resolution setting up the organization clearly specified exactly how many of each group (developed, developing and Communist) should be elected to the Trade and Development Board (of 55 members) to ensure a reasonable balance between developed and developing, and between Communist and non-Communist states. In addition, important decisions had to be taken by a two-thirds majority. A similar system could be used within the sea-bed authority. This could ensure, for example, that if decisions were to be taken by two-thirds majority, the developed nations as a whole would have more than a third of the seats; or that a certain majority of both rich and poor separately were required. It could also be provided that the super-powers (including China) were permanently represented in the Council (such a convention already applies in most UN bodies). The difficulty of this solution is that, whereas a balance between developed and developing nations could be achieved in this way, it might not secure the necessary balance between large and small, coastal and non-coastal, capitalist and socialist states, which is needed. It would institutionalize rich–poor confrontation. And it would still remain possible for major groups and powers to be outvoted.

The best solution of all is probably the simplest. Representation would be on a perfectly normal basis. The geographical groups would each elect their own representatives, as in the Security Council and other bodies; and the agency's statute would lay down the number to be elected from each group (just as the amended Charter lays down the proportions in the Security Council). Again the three super-powers should probably be permanently represented. Each group would perhaps be required to elect one non-coastal state. Thus the balance would be somewhat similar to that in the Security Council today, though the size would be twice as large (30 members). But it would also be laid down that, for all important decisions, the nature of which would be clearly specified (they would include any major amendment to the system, but not routine management functions), an overwhelming majority would be required: say, 75 or even 80 per cent. This would ensure that for any major decision of this kind to be reached, a wide majority of interests and views would have to be reconciled; but no single power, nor any small group could obstruct a decision.

Agreement on such a system would also make the related question of the balance between Assembly and Council less significant. There would of course be no object in providing a system of special voting in the Council if its decisions could be overturned by votes by simple majority in the Assembly. Major decisions would thus inevitably need to be reserved to the Council. It would be theoretically possible to provide a system under which the Assembly required the same type of voting as in the Council to reach certain types of decisions—for example, on long-term policy. The executive decisions of the Council should not however be overturned by the Assembly, even using the same system of voting, as there could obviously be no stable régime if the award of a site to an operator or a decision on the scale of royalties could be later overturned by a new vote in a different body.

Otherwise, the division of labour between the two is not seriously in dispute. The Assembly would approve the budget, elect the Director-General, elect the Council and discuss general policy. It could from time to time amend the Convention itself or parts of it (once again, at least a 75 per cent majority would be necessary). The Council would approve the issue of licences, determine the scale of royalties (when a change was necessary), decide on the allocation of revenues, lay down detailed regulations relating to exploitation, supervise the inspection system, propose the budget and oversee the finances of the organization, take emergency action in case of a threat to the marine environment, supervise technical assistance and research operations within the agency, and submit reports to the Assembly. No doubt there might be some marginal questions on which, for example, the Assembly could make 'recommendations' to the Council (or vice versa). But once the voting system had been agreed, the division of responsibility between Council and Assembly should not cause too much difficulty.

There would also be need for a number of subsidiary bodies. There have been proposals (in the U.S. draft treaty, for example) for an International Sea-bed Boundary Commission. This would consist of experts, with the necessary oceanological expertise, to review the delimitation of sea-bed boundaries. This would perhaps be done initially by the inspection service. Or member states might themselves make initial proposals, based on whatever criterion was adopted: the Commission would then examine the proposed limits and might recommend adjustments where necessary. If negotiations failed, there might be appeal to the International Court, or the sea-bed tribunal. There might also need to be some visible demarcation on the sea-bed itself. Such a commission might also be able to decide disputes between nations concerning the *lateral* boundary of the continental shelf (that between states).

Secondly, there is need for some body for overseeing the distribution of revenues. The Tanzanian draft treaty, for example, proposed a Distribution Agency for this purpose. Here the main difficulty concerns the principles which would be applied, not the body to perform it. The general principles would no doubt have to be laid down in the treaty itself. The fairest and most logical system would be something on the lines of that proposed by Tanzania: the distribution of revenues to all participating members in inverse proportion to their contribution to the UN—or, more accurately, to their contribution per head. This system manages to combine a measure of special assistance to the developing countries with the principle that the resources are the joint property of all, from which every nation, even the richest, should benefit.

A more sophisticated system of distribution was suggested in one of the Secretariat papers for the Committee.[1] The paper made the purely hypothetical assumption that by the end of the decade revenues of $500 million could be available. It then looked at the distribution of this sum among states, taking account both of population and income a head, related to each other according to five different criteria. The formulae used were somewhat complicated, and it is not necessary to describe them here. They differed not only in their degree of progressiveness, but in the extent to which they favoured very small countries or those coming near or above the middle of the income scale. A country such as Somalia, with a very low income per head, which would have received only $385,000 under a distribution based on population alone, might receive anything between $529,000 and $969,000, according to the criterion chosen.[2] As this shows, even under the most progressive scheme, the amount received by any one country would be very small (under this assumption of revenues): the total would be well under 1 per cent of the developing countries' estimated export earnings by 1980. The paper therefore suggested an alternative scheme under which the revenues were concentrated among the 25 'least developed' countries (under the current UN definition of these): this would mean that Somalia would have acquired nearly $10 million, the Sudan $53 million and Ethiopia $87 million, quite worth-while sums for such countries to receive.

A part of the revenues, as has been suggested, could be devoted to activities related to the sea-bed itself—research institutes and sea parks, for example; and another part to securing the transfer of technology to developing countries. There would still be need for decisions on the exact form expenditure should take: here too, a commission might prove useful in examining the alternatives and making recommendations.[3] The ultimate decisions

[1] A/AC138/38 of June 15, 1971.

[2] The share of the United States, which would have been 5.8% on the basis of population alone, varied between 0.22% and 1.9%.

[3] A part might also be required, according to some governments, for the administrative costs of the authority. These would be considerable and in the early years much greater than its revenues. But there is no real reason why they should not be met, like those of the specialized agencies, by direct contributions from governments rather than from sea-bed revenues, at least at first.

would no doubt be taken by the Council itself, but a commission, or a subcommittee of the Council, might serve to make the initial study and recommendations. It could also, like the Committee on Contributions in the UN, study the exact application of the formula chosen for distribution to governments, perhaps adjusting the UN scale for differences in membership, and considering any special hardship cases for which allowance might need to be made.

Thirdly, there is need for a body to consider what action is required to avoid disruption of the markets of traditional producers and to secure orderly marketing. Here again, the first decision is what type of measure would be required. A Secretariat paper on the 'Possible Impact of Sea-bed Mineral Production in the area beyond National Jurisdiction on World Markets',[1] concluded that deep-sea production of oil would have little impact on markets, that the markets for nickel and cobalt would not be seriously disrupted, and that the market for copper (of which a smaller proportion would be obtained from the nodules) would not suffer unless there was a technical break-through increasing sea-bed production. Even in the case of manganese, the most vulnerable market, the new production might be absorbed, it was felt, by an increase in demand. But, even on these somewhat optimistic assumptions, it was accepted that some regulation might be required. It was held that sea-bed exploitation should be neither subsidised nor discriminated against; a levy on each ton of metal produced could be used to represent the equivalent of the tax burden for land production. In addition, there might be some kind of compensatory arrangements for land producers affected, financed either direct from the international machinery's revenue, or from a levy on consumers benefiting from the lower prices of sea-bed products.

However, it is by no means certain that arguments for treating land and sea-bed production on an exactly equal footing will finally prevail; nor even certain that they should. Arguments that are economically rational are not always socially rational. And though there should not perhaps, in pure economic theory, be any distinction between one source of production and another, in political and social reality there often is. Given that the livelihood of hundreds of thousands may be at stake, and that the

[1] A/AC138/36.

foreign exchange earnings and revenues of a number of developing countries could be drastically affected, it is not self-evident that, in a world where coal mines have long been artificially protected against oil and gas in richer countries to safeguard employment levels, miners of copper and manganese, in countries where unemployment is a far larger problem, should not receive some protection against capital-intensive sea-bed production. There is a case for saying that there *should* be, at least temporarily, some discrimination, or a very generous compensation scheme, to assist the process of adjustment. A commodity scheme which made no distinction at all between the two different sources of supply would not be an adequate alternative to assistance of this kind. The main outlines of such a system would need to be set out in the treaty establishing the régime, and re-examined from time to time. Whatever the principles finally decided upon, there is likely to be a need for some permanent body, whether separate commission or subcommittee of the council, to supervise its application in practice.

There might also be need for a further intergovernmental body for supervising sea-bed operations and exercising day-to-day control of exploitation. Under the U.S. treaty, an Operations Commission was to issue licences for the international zone, supervise the operations of licensees, resolve disputes concerning operations, initiate proceedings for violations of the convention, collect fees and other payments, arrange the dissemination of information and so on. Some of these duties could perhaps be equally effectively performed by the Secretariat, or by the Council, or both, without any separate body. But there might be a need for some specialized body in continuous session to reach decisions relatively rapidly on licensing, and to lay down general rules and technical regulations.[1] A permanent body, with representatives having the necessary technical qualifications but also geographically balanced, like other 'expert committees' in the UN, might perform the function required here.

Then there is the task of organizing the transfer of technology

[1] The U.S. treaty proposed a separate commission on Rules and Recommended Practices, composed of people of technical qualifications and experience to lay down the technical regulations governing operations within the international area. It seems doubtful if a separate body for this is necessary. The Maltese treaty proposed an Ocean Management and Development Commission, a Scientific and Technical Commission and a Legal Commission.

and assistance in sea-bed knowledge and skills. Given the imbalance which at present exists here, this would be an essential task. It might take a number of forms: imposing obligations on sea-bed operators to employ and train a fixed proportion of developing countries' personnel; the organization of training programmes in UN institutes, perhaps set up for the purpose; fellowships and training in developed countries; the dispatch of experts to developing countries to undertake university teaching and technical training there; the establishment of regional institutes; the dissemination of information and specialized publications; organizing international oceanological expeditions; securing publication of the results from all exploration work, commercial and scientific; and the organization of data banks. Co-operation with other organizations, such as UNESCO and the IOC, in these tasks would be needed. Here too, in view of the scale the operation would eventually assume, there is something to be said for the creation of a special committee which would be permanently charged with overseeing the work.

Next, a corps of inspectors, technically qualified but geographically balanced, would be required. Their task would be comparable to that performed by national agencies responsible for the inspection of installations within continental shelves. They would need to ensure that the detailed operational standards laid down were being strictly complied with. These would include technical, safety and environmental standards. The inspectors would be able to make unheralded visits and spot checks. They would issue reports on each visit to the Council or the responsible committee.

But they could be used for other purposes as well. They could help in supervising the delimitation and demarcation of sea-bed boundaries. They could help to verify the demilitarization of the sea-bed. They could examine charges of interference with other uses, actual or threatened. And if some kind of supervision of research activities, even purely scientific, were finally decided on, as some countries demand, it could have a role in this field. All operators would need to give an assurance that they would allow the inspectors unimpeded access to their installations at all times. Against this, they in turn would receive assurances that all information obtained would be treated confidentially, at least for a certain number of years. The inspectorate might require expensive equipment, in the form of submersibles, underwater photography equipment, television cameras, telecommunications equipment, even, possibly, aircraft.

Lastly, it is generally recognized that there would need to be some institution for the settlement of disputes. Disputes would be mainly of two kinds: between enterprises, and between an enterprise and the authority. Occasionally there could also be disputes between a member state and the authority, or against some other member, concerning the interpretation of the treaty. In some cases the agency might need to take legal action against a company for non-fulfilment of obligations, or damage to the marine environment. In all these cases, some special sea-bed tribunal would be required to hear the disputes. If necessary, there could be a final appeal to the International Court of Justice on the interpretation of the treaty itself.

Obviously, the main legal principles to be applied, including those relating to liability, would be laid down in the treaty itself. In particular, this would need to make clear whether a system of strict liability (that is, regardless of proof of negligence) would be applied; and whether a maximum figure of damages for any one accident should be laid down (as was done in the recent Convention concerning liability for damage to a coastal state from oil pollution resulting from an accident). Through subsequent Conventions and regulations, a new law of the sea-bed would gradually be built up.

If allocation is made direct to operators rather than to states, under what law will the individuals and firms involved operate? The rules relating to exploitation will of course be contained in the treaty itself and its annexes, and in the contracts into which individual operators enter; and these will be enforced by the authority. So far as criminal law and some aspects of civil law are concerned, however, this will obviously not be the case. Here probably the best solution is that already adopted in another international area, Antarctica: every exploiting enterprise, and its personnel, would operate under the criminal and civil law of its own country, or the country in which it was incorporated. This would be something like the flagstate jurisdiction that exists at sea in relation to oil rigs. This could not, however, cover the case of exploitation direct by an international organization, whether regional or international. Under these circumstances there would require to be some prior agreement as to the criminal jurisdiction which would apply to operations of this kind.

Even if an institutional structure roughly on these lines was agreed, there would remain a general question: how much of the detail would need to be specified precisely in the treaty, and how much could be left to the institutions to decide at a later stage? Here, as over voting powers, there can be not the slightest doubt that some of the larger powers, especially the Soviet Union, the United States and China, simply will not enter a system in which a large range of discretionary power to alter the system is accorded to the institutions set up, however they may be constituted. Under any imaginable treaty it would be impossible to foresee and allow for every change in circumstances that could occur; and procedures for revision must therefore be provided for. But a considerable degree of detail will need to be laid down in the treaty in advance, if such states are to be persuaded to accept the system. One way of reconciling the two needs—a clearly defined and agreed system, and a method of making adjustments where necessary—is that used in the specialized agencies such as the International Telecommunication Union, the Universal Postal Union and the World Meteorological Organization. These bodies nowadays have two or more basic instruments: a convention under which the agency is constituted, setting out its basic structure and powers; and a set of regulations under which the detailed rules and procedures are laid out. Whereas the former remains relatively constant, the latter is reviewed regularly: usually at every meeting of the Assembly (that is, every 3 to 5 years). Some of these organizations employ a system under which at least some of the new regulations passed in this way come into effect automatically for every member unless it issues a reservation within a certain period. A somewhat similar system might be used for amending the less important regulations, which might be contained perhaps in annexes, that could be regularly revised. Amendments to more important features of the system, contained in the treaty itself, would require a bigger majority, probably of at least three-quarters. They could not however come into effect only for those who ratified them: there can be only one system, for example, for dividing up sea-bed blocks or distributing revenue. Decisions by the approved method would become binding for all.

There have been a number of proposals for regional bodies to take over important sea-bed functions. Some of the inspection work, for example, could be undertaken by regions, provided that the standards applied were uniform. Training and technical assistance work could be arranged through regional headquarters (as in the specialized agencies). Regional research institutes might be established. If these were the only functions, and uniform standards were maintained, such bodies would be in effect merely branches of the main authority and so quite harmless. The main motive for regional régimes, however, seems to be to allow countries of particular regions to maintain wider limits—or to apply different principles—than in the régime as a whole. And this could be conceded only at the expense of depriving the régime, and therefore countries of other regions, of part of the revenues which they might otherwise legitimately expect.

There is another danger in allowing regional bodies too large a role. The essential transfer of skill and wealth that the system is supposed to bring about could be inhibited. The North American and European regions would, under any scheme, inevitably be those where the most financial resources, technological skills and entrepreneurial techniques would become available; the Asian, African and Latin-American regions would be those where, for political or other reasons, they would follow least readily. Autonomy for the regions, far from benefiting the latter regions, therefore, could severely impoverish them. There is a good case for sensible devolution to local bodies of particular functions to which they are suited; but to give them too wide a role in order to appease regional pride would be not merely to destroy the essentially international character of the régime, but to slow down for many decades sea-bed exploitation in precisely those areas which have most need of it.

Another institutional problem concerns the agency's relationship with the UN and the specialized agencies. The sea-bed authority could clearly not be a subordinate body of the UN, in the way that the UN Conference on Trade and Development (UNCTAD) and the UN Industrial Development Organization (UNIDO), or the existing Sea-bed Committee for example are. Among other things this would nullify the effect of any special voting procedure established, as all those bodies can in the final resort be given direct orders by the UN. Even the status of a specialized agency might be too subordinate. Though in practice virtually independent, the agencies are obliged to report regularly to the UN Economic and Social Council (ECOSOC), to have their budgets scrutinized and their operations inspected by the Joint Inspection

Unit, none of which would seem appropriate in the case of the sea-bed authority. On the other hand, such an authority would require to enter into the closest co-operation with the UN and especially with agencies such as IMCO, WMO, and UNESCO, which are all in some way concerned with the oceans. And there should be some machinery, possibly through the Administrative Committee on Co-ordination (ACC) and the General Assembly, or a World Oceans Council (see page 227 above), by which the necessary co-ordination could be achieved. Its status might therefore be most comparable to that of the World Bank and the International Monetary Fund, which are members of the UN family and are represented in the ACC, yet are relieved of some of the duties and obligations of the other agencies and so are considerably more independent.

Finally, there is the question of the legal status which the agency should enjoy. It would clearly require full legal personality, to be able to enter into contracts, to make purchases, to sue and to be sued. In this respect its position would not, even though it was largely a commercial undertaking, be essentially very different from that of the UN and the agencies. Article 104 of the UN Charter lays down that 'the Organization shall enjoy in the territory of each of its members, such legal capacity as may be necessary for the exercise of its functions and the fulfilment of its purpose'. The Convention on the Privileges and Immunities of the UN provides, in Article 1, that

> the U.N. shall possess juridicial personality. It shall have the capacity:
> (a) to contract
> (b) to acquire and dispose of immovable and movable property
> (c) to institute legal proceedings.

Most of the specialized agencies have the same or similar provisions in their constitutional instruments. The sea-bed agency would need powers at least as great as these. It would also require roughly the same diplomatic privileges and immunities for its employees.

If this, roughly, is the form which the organization would take, how could it enforce the system? Its main sanction would be the power to order the forfeiture of an area, or exploiting rights, against any state or enterprise that violated the rules. In most cases this would probably be sufficient to ensure compliance by any party that remained within the system. The real problem would be in dealing with nations or companies that either refused to join the system in the first place, or that joined and then left because they rejected decisions reached against their interests. This is a far more difficult question, which we shall consider again later (pages 296 to 299 below). For the moment, it need only be said that such situations would have to be clearly foreseen in advance, and the sanctions to be applied—such as the refusal of all members to purchase materials acquired through illegal exploitation—would need to be clearly laid down in the treaty in advance, so that they were applied almost automatically. All experience of the use of sanctions suggests that only if they are applied vigorously and universally from the beginning are they likely to exercise much effect. Any hesitation, or any attempt to leave to the decision of individual governments what measures they should apply, would only ensure that the sanctions were wholly ineffective—and the entire régime could easily collapse in consequence. Once it was known that one country could defy its provisions with impunity, there would be little incentive left to the remaining members to apply them effectively.

The Limits

But, even if agreement could be reached on the system for exploitation, and the powers and institutions of the authority to administer it, there would remain a further problem, more difficult than all—the problem we have many times encountered in this study, and perhaps the most important of any which surrounds the question: where should the limit between the national and the international area lie?

Before looking at the individual problems, it may be well to examine the general principles that will have to be applied if a satisfactory solution is to be found. As we have seen, at the root of many of the difficulties over the sea-bed issue is a fundamental conflict of interest between coastal and non-coastal states: in other words, between particular interests and the interest of the international community.[1] In other cases, there is a conflict between other special groups—naval powers, developing fishing states, distant-waters fishing countries, developed countries able to

[1] According to the Netherlands delegate to the Sea-bed Committee, about two thirds of the international community are 'primarily' non-coastal, including 30 land-locked, 18 shelf-locked and 40 states with short coastlines.

exploit sea-bed resources and so on. In every case, therefore, what is required is a *balance of interests* among these various groups—and especially between the coastal and non-coastal.

This leads to certain general conclusions. If a solution is to be found, it cannot be simply by drawing fresh lines on the map, giving new authority to one set of states and therefore reducing the freedom of action of others. It must be by finding new ways of *sharing* authority, between individual states and the international community as a whole. Thus, for example, if because of changing technology coastal states are to be granted extended jurisdiction for certain purposes—say, pollution control, fishing, or sea-bed resources—this can only be in return for accepting new responsibilities as well. These responsibilities will be of three main kinds: the observance of certain internationally defined principles; the recognition of the powers and activity of international authorities; and the obligation to pay a proportion of the revenues which they or their nationals acquire as international revenues.

Thus, for example, if coastal states are granted new authority to manage and conserve fisheries off their coasts, this should be only in accordance with principles that have been internationally agreed. If they are given new powers to take action on pollution in the seas, this should be only if they also acknowledge the powers of international authorities to lay down rules and supervise the maintenance of environmental standards in the same areas. If they are to be given the right to control sea-bed exploitation in areas miles from their own coast, it should be only in return for the payment of royalties to the international community for the revenues so acquired. International principles, international authorities and international revenues would reflect the international interest in areas that are not the exclusive concern of any single state.

Thus, as suggested earlier, it may be more sensible no longer to think in terms of a rigid barrier between one zone that is national and another that is international. In practice, even in the most national area of all, territorial waters, some international obligations are already recognized, and will be increasingly required (for example, in relation to pollution). Conversely, even in the most international area, the high seas, some rights for individual governments or companies may need to be accepted (for example, in relation to fishing or exploitation). By recognizing a balance of obligations and rights in all areas, it is possible to reduce the importance of any single limit, and to match each one more accurately to the purposes it is designed to fulfil.

Perhaps the simplest way of considering the question is to look first at the areas closest to the coast and then to move outwards.

Within the territorial sea, many of the rules laid down in the 1958 Convention on that question will no doubt be reaffirmed. The area will remain under the sovereignty of the coastal state and in all essentials a part of its territory. That state will be able to exercise full authority there, even if, in practice, as today, offences committed in foreign ships by foreign nationals are often left to be dealt with under the jurisdiction of the flag state, either by the master of the ship or by a foreign consul. It is quite likely that many of the articles of the 1958 Convention on this subject will scarcely need to be changed at all.

The really contentious point here concerns the right of 'innocent passage'. If a wider limit for the territorial sea is agreed, it becomes the more important to spell out this right more precisely. This would involve providing a definition of 'innocent' and so specifying the grounds, related to security, pollution or other matters, that alone would justify interference. An international tribunal to hear disputes on the question might be needed: at present the coastal state itself can judge what passage is 'innocent'.

The principal point in dispute here relates to the rights of warships (see above, page 157). So far as the territorial sea generally is concerned, it would no longer seem reasonable to maintain that warships of all nationalities should have an automatic right to sail as close to the coasts of another state as they wish, whenever they please. The real crux of the matter concerns the definition of the word 'innocent', and it is not very logical to say that a warship's passage remains innocent until the moment it begins to fire its guns. Outside international straits, therefore, there would seem to be good grounds to demand prior *authorization* for such ships to sail in these waters. But it would seem essential to make a distinction between recognized sea lanes, especially those leading to or from international straits, and other territorial waters. In the former, warships might be subject only to prior *notification* (not authorization). Unless such freedom were allowed, the power to deny passage to warships could be used by a strategically placed country to give overwhelming strategic advantages to one country or block at the expense of another. Provided that this were agreed, and the definition of 'innocent passage' tightened to prevent arbitrary action, it is less certain that a special régime of 'free transit' is required as well.

A more important point concerns the outer limit of the territorial sea. As we saw earlier, there is probably now an overwhelming majority of states which will favour a twelve-mile limit for the territorial sea. This will include a number which would themselves prefer a smaller limit, but which would accept a twelve-miles maximum, for the sake of general agreement. There will certainly be more than enough to provide the two-thirds majority required to be carried at the Conference. It would then be widely accepted as representing currently accepted international law on the subject: for example, by the International Court.

But some of those who would accept this will seek to secure concessions in return. They will seek to link the questions of limits and the régime. They will recall that the Conference was to reach decisions on limits only 'in the light of' the régime. And they may seek to ensure that any limits which are laid down are contained in the general treaty relating to the sea-bed régime, rather than in a separate convention, so one cannot be agreed without the other. Thus the rich countries may find that they can secure agreement on a twelve-mile territorial sea only in return for concessions on other points: perhaps a wide 'economic zone' or a strong régime.

Assuming that a twelve-mile limit for the territorial sea is agreed, a new problem would arise: should there continue to be a contiguous zone beyond that line? If changes in the speed of communication, in military technology and in economic and scientific conditions have brought the need for a wider territorial sea, they have also, it could be held, brought the need for a wider contiguous zone as well. And as such a zone has been demanded above all on security grounds (though this is not one of the purposes admitted in the 1958 Convention), it is possible that some such extension will be widely demanded. There is obviously some danger in accepting a perpetually creeping jurisdiction of this kind. If a new limit for the contiguous zone is admitted now, will this soon become the limit demanded for the territorial sea? It is doubtful whether a zone of more than twelve miles is really required for sanitary, fiscal and immigration purposes. But if any contiguous zone of this kind is to be created at all, it should not be more than a further six miles beyond the territorial sea—making eighteen in all; or, better, twelve miles for both zones combined.

There are a number of other questions that would arise indirectly from any agreement which might be reached concerning the breadth of the territorial sea. Special provision would have to be made for the case of archipelagos. It must be accepted that there exists a genuine problem here. Although the extreme claims—to be able to draw lines from the outermost points of all islands of a group, however dispersed—could not be accepted, it is undeniable that states formed of island groups do confront genuine problems, especially in the field of security. But a compromise is not impossible. For the territorial sea itself, the normal limits and the normal method of measuring base-lines should be used: with, possibly, a maximum distance of 48 (instead of 24) miles between islands which could be linked in a single stretch of territorial waters. For the waters between the islands, a new form of intermediate zone could be created, or the rights normally enjoyed within the contiguous zone might be acknowledged. These could include the right to demand notification of the passage of warships within such an area, but should not include any special fishing or other economic rights beyond the generally accepted limit.

Another difficult question relates to the rights of land-locked states. This arises quite independently of any sea-bed régime. It is probable that the land-locked countries will seek at the Conference some strengthening of the present international rules concerning access to the sea (as they attempted to do at the time of the last conference on the law of the sea in 1958). Article 3 of the Geneva Convention on the High Seas provides that states having no sea coast 'should' have free access to the sea, and that it should receive from states lying between it and the sea, 'by common agreement' and 'on a basis of reciprocity', free transit through its territory; and for ships flying its flag 'treatment equal to that accorded to' the ships of the state itself, as regards access to seaports and the use of such ports. Moreover 'states situated between the sea and a state having no sea-coast should settle, by mutual agreement with the latter, all matters relating to freedom of transit and equal treatment in ports in the event of such states not being already parties to existing international conventions'. Since these obligations depend on 'mutual agreement' and a 'basis of reciprocity' (where no genuine reciprocity of situation exists)—not to speak of ratification of the Convention by neighbouring states—they do not really provide any very firm assurances to the land-locked states; and some of these, such as Bolivia, have continued to suffer difficulties from unco-operative neighbours. There is a more specialized Convention on the Transit Trade of Land-locked States, but only 23 states have ratified it; and most of these are themselves land-locked states. At present, therefore, land-locked states in general remain dependent on the goodwill of neighbouring states. And transport facilities, railways, roads and airports which may be essential to their standard of living may be allowed to go unrepaired, or even unbuilt, without their enjoying any right of redress. There is a good case for providing better assurances on these points, as well as rights to participate in sea-bed exploitation, for these countries, often among the poorest of the world.

Another question that will have to be settled relates to islands, including artificial islands, and whether or not these can create their own shelves or economic zones. Here, a definition of a qualifying island could be based on the size of the island; whether or not it was populated; on the distance from the metropolitan country; or some combination of these. A particularly difficult problem concerns colonial islands. These presumably would be treated as independent entities, able to create their own shelves on the same terms as other islands. But the question would then arise whether any provision was necessary to ensure that the resources of such shelves go to the inhabitants of the islands and would not be those of the metropolitan power. These are important points, which would certainly require solution in establishing a régime.

So far we have envisaged on the surface of the waters a limit of twelve miles for the territorial sea, with a possible contiguous zone of not more than six miles beyond, together with special arrangements covering the freedom of passage through straits and for archipelagos. Next, as we saw, there is a demand for special rights in relation to pollution; and Canada has already asserted such rights. This undoubtedly becomes an increasingly serious problem with ever larger oil tankers and a greater volume of shipping. But the fact that strict measures should be taken, here as elsewhere, does not necessarily mean that they should be taken on a unilateral basis. Again it is a sharing of authority with international bodies that is needed. International legislation can provide general rules which coastal states might be given authority to enforce. Provided that the international system is adequate, national measures against shipping are not only unnecessary but undesirable. For they could bring about a confusion of divergent standards for different areas, which it would be almost impossible for individual shipowners to conform with.

Indeed, where international rules exist, national measures in this field can represent a form of arrogance: 'you must adopt our standards, not those which the rest of the world has accepted'; or even, 'we must apply higher standards than anybody else, because our coasts are more important'. Some standards, for example general rules of navigation, standards of ship construction and others, clearly cannot vary widely from one country to another; and here the right of individual states to demand their own must be resisted. Only where they are directed purely to local conditions —for example, rules of navigation in particular areas (such as Arctic channels), demands for advance notification of voyages in particular places—should such unilateral controls be accepted. Even here, their form should be negotiated, not imposed. If, therefore, new and special rights of jurisdiction are to be accorded to coastal states in this field, it is important that they should be internationally defined, and not seized unilaterally. The coastal state would act on behalf of the international community and according to principles it laid down.

Comparable problems occur over fishing management. Here too, control should be *shared* between national and international bodies. In most cases the essential unit to be considered is not a particular strip of water, or even a series of strips, but a particular stock of fish. This stock may extend over wide areas that are universally recognized as being international, as well as within coastal waters. What is required is the sound management, conservation and harvesting of that stock as a whole. This can only be done on an international basis. The fisheries commissions, under the Fisheries Committee of the FAO, are the best bodies to undertake this task. They may allow the coastal states to take certain actions, according to agreed principles, in their own areas. As in the case of pollution, therefore, the most essential step to be taken here is not a simple extension of national jurisdiction, but the development of the kinds of international bodies and rules which alone can do the job required.

But even if this is accepted, it does not solve the problem of fishing limits. International bodies must recognize some special rights for coastal waters. Whatever the economic arguments, some extension of fishing limits now seems to be inevitable. The development of long-range fishing fleets equipped with the most modern and expensive fishing equipment, and capable of encroaching on waters traditionally fished by coastal states, has created a wholly novel situation. Even distant-water fishing countries, such as the United States and the Soviet Union, which have the greatest interest in narrow limits, have now explicitly conceded that some adjustment is required.[1] The question is simply: what should be the rights conceded, in what areas and in return for what obligations?

Some of the issues which arise here are not unlike those which occur over the resources of the sea-bed. Particular countries, favoured by geography in possessing important resources in the waters off their coasts, believe that they have a special right to benefit from this bounty of nature. Countries of developed techniques, on the other hand, regard the resources as common, in the sense of commonly available, so that all should be able to exploit them on a basis of 'first come, first served'. Finally, non-coastal states, incapable of fishing themselves, may regard them as common in a different sense: common property from which they should draw some benefit. Each of these positions has a certain natural logic about it. Each of them reflects a different view of 'rights' in marine resources, all equally reasonable according to their own premises. And to decide where the boundaries should lie on their basis can only be a matter of arbitrary judgement.

It is arguable that there could and should be no uniform limit of fishing rights, that this should be decided by the appropriate commission, that it should depend on the stock of fish concerned, its migrating habits and its abundance and even on the capacities of the fishing fleet of the coastal state or region in question. Again, varying limits could be held necessary on the grounds of the differing *needs* of coastal states for protection. Iceland, as a country almost entirely dependent on fishing (90 per cent of her exports are fish products) could claim to have a greater need for exclusive rights than the United States, which not only is the

[1] In the last year or two, a number of rich countries as well as poor have come out in favour of wide limits for fishing and other economic rights. Australia and New Zealand say that they will claim 200 miles for fishing. Canada supports 200 miles or the edge of the continental shelf, whichever is greater. Norway supports a limit for the shelf of 200 miles or 600 metres in depth at the choice of the coastal state. Thirty or so nations have now extended their fishing limits beyond 12 miles. Many other countries have accepted the principle of extensive coastal state jurisdiction.

richest country in the world but whose fleets are highly efficient and well equipped, and could stand up to a considerable degree of foreign competition. Thus it could be argued that what is required here is a system of *preference* from developed to developing countries, similar to those recently granted in the field of trade. Developing countries or continents would be given special preferential rights in areas close to their own coasts, but no corresponding rights would be given to developed states off theirs.

In practice, it is perhaps unlikely that a system of limits varying according to stocks or fishing capacity will be adopted. Law tends to prefer uniform solutions. So does international politics. Many coastal states demand that there should be a zone beyond the territorial sea where the coastal state can *always* claim exclusive fishing rights and impose conservation measures. Other countries are beginning to accept this need. The question is mainly where the limit should be drawn.

If, as it seems, a wide 'economic zone' for the coastal state is conceded, this will certainly include fishing rights as well as sea-bed rights in that area. The breadth most often claimed for this is 200 miles. If local fishermen cannot exhaust the stock, however, this may not be accepted by the large maritime states, especially the United States and the Soviet Union. The land-locked will also resist such a massive encroachment on their natural heritage. There may be a compromise at, say, 100 miles. More likely 200 miles will be accepted but on condition that extensive *international* obligations too are recognized within the zone, including the obligation to pay part of the revenues acquired to the régime. Alternatively, there could be a kind of regional fishing zone in which the coastal state would at least have to share fishing rights with other countries of the same region. Such a scheme would distinguish sharply between competition from fishermen of the same region, sharing common rights in closely neighbouring stocks, and that from long-distance fleets of highly industrialized countries from far away, which can more reasonably be regarded as encroaching on the resources of others. All of these are possible compromises which will be bitterly debated within the Conference.[1]

Whether there is a separate system for fishing or a general

[1] On anadromous species, such as the salmon, Ireland has proposed that the spawning nation should be given sole control; whereas Japan takes the opposite view, favouring free fishing in the oceans.

economic zone, what should be the limit of national jurisdiction over sea-bed resources? This is the final question that arises on limits, and is perhaps the most important of all.

The first point to be determined here is whether the limit should be laid down in terms of depth or of distance. There is today increasing recognition of the advantages of a limit expressed primarily in terms of distance from the coast. A definition based on depth is difficult to measure accurately, or to demarcate in cases of dispute. The sea-bed does not always follow an exactly even incline; and in many cases there might be many different areas, at roughly the same depth, each of which could be claimed to represent the final boundary. Or there could be deep trenches intersecting a shallow area, which could be the subject of bitter dispute. A limit based on distance, on the other hand, would be relatively easily defined and demarcated, both on the surface and on the sea-bed once the necessary baselines had been chosen. It would be more easily comparable with the other limits of jurisdiction used: for the territorial sea, the contiguous zone and so on. It would be a limit that was far more readily apprehended and understood. Finally, such a basis would be generally in conformity with the principles of natural justice: roughly equal areas would be given for a given length of coastline (even under this definition the country with a shallow shelf would retain the advantages of easier exploitability). It is more obviously fair that a nation has a right to resources within a certain distance of its shores than that it has a right to resources in the area that is a 'prolongation' of its own land mass, even if these are 600 miles away.

For all these reasons, it is better to make distance the primary criterion employed. But, as the depth criterion has been widely used since 1958, and as a number of countries have already begun exploitation on this basis, it would have to remain open to any nation to claim all areas of a depth less than 200 metres off their coasts. This would anyway be necessary in practice where exploitation had already begun in such areas. And this in turn would make it essential to accord similar rights to countries which had not begun exploitation (otherwise there would simply be a rush to begin it in the period before the Convention was ratified).

If some combination of distance and depth of this kind was accepted, what should the distance be? The fact that many coastal states are today demanding a wide 'economic zone', of course reflects their own interests. There is a case perhaps for some control by the coastal state in wide areas off its coast. But the demand for exclusive jurisdiction and ownership in a 200-mile zone could partly nullify the whole concept of the 'heritage of mankind', since it would place a very large proportion of the easily exploited resources, and so of the available wealth of the sea-bed, in the hands of coastal states. Nor does such a limit have any genuine basis in geographical reality. It would not approximate to any known geographical concept. It would mean that coastal nations were claiming huge areas far beyond sight of their own coasts, with which they had no greater contact or relationship than the peoples of other nations. Some countries could claim larger areas of the sea-bed than they had on land. Coastal states, which are already fortunately endowed in relation to the land-locked in having direct access to the sea, would appropriate all the most valuable areas of the oceans at the expense of the non-coastal states. And those with especially long coastlines, such as the United States, Canada, South Africa, Australia, the Soviet Union and others, would acquire especially large benefits in relation to smaller countries. The greater proportion of the resources of the sea-bed, which can reasonably be regarded as the equal property of all peoples, would be taken over by a small minority of states having special geographical advantages.

Any new economic limit would clearly have to be at least as wide as the present minimum; that is, wider than the 200-metre depth line. Indeed, it must probably go beyond this point. The limit of 40 miles, which has been proposed by some land-locked and other countries, is almost certainly too narrow to be practicable. A possible limit might thus be a line 50 miles from the baselines on the coast. This was the limit which was calculated by the Commission to Study the Organization of Peace in the United States to be roughly equal to the average of the physical shelf for the world as a whole. This in itself would mean that it would correspond, in a rough and ready way, with a physical reality, and not be simply a line drawn at random, as a crude but arbitrary political bargain. It would at the same time conform with traditional international law, which has always sought to relate its definitions to the concept of the physical shelf. It would correspond, more accurately than any other distance, with the judgement of the International Court, concerning the 'prolongation' of the land mass: the physical shelf is the only clearly defined 'prolongation'. It would give a conveniently round number. It would define the word 'adjacent' in the 1958 Convention. Above all, it would represent a reasonable compromise between the extreme claims of the wide-limit countries and those of their opponents. It would give a substantial strip of territory to every coastal state beyond its shores: nearly 10 per cent of the whole sea-bed area would be reserved to these fortunate states, and a far higher proportion of the easily exploitable area (about half the slope and over a third of the continental margin). Yet at the same time a reasonable share of the easily exploitable area would still be reserved for the régime itself.

But the basic aim—to balance the interests of coastal and non-coastal states—could be met in quite a different way. Instead of establishing that balance by the drawing of a line on the map, it would be achieved by the *sharing* of jurisdiction and benefits within a wider area.

This could be done by a modified form of economic zone. If coastal states were to be given control over resources far into the seas and far beyond the physical shelf, they would have in return to accept obligations in that area as well. These would include the duty to apply international standards and rules, for example on pollution, fish conservation, safety, free navigation and other matters; and the obligation to pay part of the revenues gained from the resources for international purposes. This therefore would have something in common with both the U.S. trusteeship idea and the Maltese plan. The basic feature of such schemes is that in the intermediate zone authority and revenues are shared between the coastal state and the international community. The system provides considerable advantages to the coastal state—in terms of control over exploitation off its coasts, and greater revenues from it. But in return it gives up some income to the international régime. Its great merit is that it softens the distinction between the area exclusively controlled by the coastal state and the fully international area beyond. It thus satisfies some of the natural appetites of the coastal states to control resources off their shores and to derive substantial economic benefits from them; yet at the same time secures a relatively large area—larger than is likely to be obtainable by any other plan—from which the international régime secures revenues.

There are two other compromise schemes which are worth mentioning. One would be a system of *temporary* trusteeship.

What this would mean is that, beyond an area of full national economic rights—say, 200 metres in depth as in the U.S. proposals—the coastal state would, in a further area beyond (say, to 500 metres depth, or 200 miles), be accorded additional rights, but only for a limited period. These might include the right to choose the operators, or to draw revenues, or any others which might be agreed on. After a period of, say, 15 or 20 years, the area would revert to the international system and be exploited on the same principles as the rest of the area. This, again, is a compromise between the demands of the coastal states and those of the non-coastal. The system would have the disadvantage that it might stimulate the government concerned to undertake the maximum exploitation during its limited period of control, and so could lead to wasteful or damaging use of resources, without sufficient control or good resource management. This might be remedied by a provision for agreement between the coastal state and the régime on any exploitation contracts which it entered into.

The second variation would be a system under which payments for the régime were made not only for the economic zone but for the existing continental shelf, and even in territorial waters. Thus the price to the coastal state of taking over the trusteeship area or economic zone would be having to make payments for these inner areas. Alternatively, there could be a system by which the coastal state's power of control over exploitation became less at each 50 miles, and its payments to the régime greater. There could even be a provision for payments in respect of fishing revenues by the coastal states within these areas. Politically, this would represent a trade-off. A coastal state would acquire some rights over a wider area than at present; but the non-coastal states would acquire revenues from some area which are at present purely national.

All these compromise schemes have the attraction of considerable logic and natural justice. Coastal states would have both rights and obligations in the sea opposite their coasts, and both in proportion to the distance from the coast. The resources of the sea, as well as the sea-bed, would be acknowledged as the natural heritage of mankind; but particular countries would be acknowledged as having special rights to that heritage in particular areas. In a sense, this conforms with the emerging reality. For the idea of a 'natural relationship between man, the land and the sea' as propounded so far, is an archaic one, based on an old-fashioned

Table 4

Distribution of Sea-bed Resources according to Various Proposals on Limits

	INTERNATIONAL AREA	COASTAL STATES
40 nautical miles from coast	Area: 346.87×10^6 sq. km	Area: 15.66×10^6 sq km
Hydrocarbons	41 per cent of total ultimate resources, including 20 billion barrels already discovered to date	59 per cent of total ultimate reserves, including 90 per cent of proved reserves
Manganese nodules	All known mine-grade deposits	No known mine-grade deposits
Other minerals	Some near-term prospects, including Red Sea metalliferous muds and brines	All immediate prospects and most in foreseeable future
200-metre isobath	Area: 340.36×10^6 sq km	Area: 21.90×10^6 sq km
Hydrocarbons	32 per cent of total ultimate resources, including some reserves discovered to date; near-term prospects on outer shelf and upper slope	68 per cent of total ultimate resources, including 167.5 billion barrels proved (almost all reserves discovered to date)
Manganese nodules	All known mine-grade deposits	No known mine-grade deposits
Other minerals	Possible near-term prospects in metalliferous muds and brines of Red Sea rift. No other prospects in foreseeable future	Immediate prospects and almost all in foreseeable future
3,000-metre isobath	Area: 318.15×10^6 sq km	Area: 45.42×10^6 sq km
Hydrocarbons	Only 7 per cent of total ultimate resources; some long-term prospects in continental rise and deeper parts of small ocean basins	93 per cent of total ultimate resources, including *all* of proved reserves and immediate prospects
Manganese nodules	All known mine-grade deposits	No known mine-grade deposits
Other minerals	No prospects in foreseeable future	All immediate prospects and all in forseeable future
200 nautical miles from coast	Area: 288.04×10^6 sq km	Area: 77.08×10^6 sq km
Hydrocarbons	Only 13 per cent of total ultimate resources; long-term prospects in continental rise	87 per cent of total ultimate resources, including *all* of proved reserves and most immediate prospects
Manganese nodules	Most known mine-grade deposits	Some mine sites adjacent to volcanic islands in the North and South Pacific

Source: UN Doc. A/AC138/87: Economic Significance of Various Limits of National Jurisdiction.

and narrow geographical view. In wider terms, one might more properly conceive of the fish off the west coast of Latin America as 'naturally' the property, not simply of Peru and Chile, but of Latin America as a whole; or even of the Americas as a whole; and finally of the entire world. This conception—of world rights in world resources—is one towards which the UN has been half groping in the sea-bed discussions during the last five years. But it needs to be made more explicit. A scheme in which rights were *shared* between coastal areas and the world as a whole, as here suggested, would implement it more fully.

There is, therefore, no lack of difficult problems in the field of sea law for the fateful conference, over and above those relating to the régime itself. They are indeed so vast, that it is highly doubtful if all the many questions to be resolved, with their complex interrelationships, will be decided for years. There may in practice be a series of conferences during which various bits of the jigsaw puzzle are gradually put in place, one by one. Though probably no single issue will be finally agreed until every other part of the jigsaw is visible, it may be possible to reach *provisional* agreement on various points, step by step, in this way. Eventually by this process, it may be possible to arrive at a system that will reconcile the aspirations and interests of deep-sea fishers and in-shore fishers, of coastal and non-coastal, of rich and poor, and large and small alike. A genuinely international system for developing world resources may for the first time be established.

CONCLUSIONS

The Challenge

A Unique Issue

There are a number of reasons why the sea-bed question has a very special—indeed, incomparable—interest as an international issue.

Its first and most obvious feature, differentiating it from all others, is its unique importance. What has been under discussion has been, quite literally, the control of two-thirds of the surface of the earth, and possibly a much higher proportion of its remaining resources.

This is the kind of question which has never before in history been debated by the representatives of governments sitting round a table, seeking to reach an agreed settlement acceptable to all. Possibly the negotiations nearest in importance in the past have been those following great conflicts, to determine the territorial dispositions and the distribution of spoils at their end: meetings such as the Congress of Vienna, the Congress of Berlin, and the Versailles Peace Conference. But even these have been of less ultimate significance in their consequences than the discussions at present proceeding about the resources of the sea-bed. For these will determine, probably permanently, the future of what may be an important part of man's available wealth, and what will certainly be a very significant area of human activity over the next century or two.

The scale of its importance has clearly influenced the tactics employed by governments, and groups of governments, within the debate. Most UN resolutions, though they are often passionately debated and discussed, and their terms violently disputed, can never be of fatal consequence to any government. They are, for the most part, no more than recommendations which can be obeyed, flouted or ignored. But the discussions on the sea-bed will have a vital, long-term effect on the economic interests of almost every member. And the issue has therefore been given far deeper and more wide-ranging consideration, both within governments and in the UN, than other UN questions. There is every indication, moreover, that the final decisions to be taken will be fought out with a passion and intensity which has never been equalled even in the UN's stormy history.

The importance of the question has been well understood by those who take part in the debate and by the governments who instruct them. But there is no evidence that it has yet been appreciated by the public as a whole. One of the most astonishing features of the discussions, given their unique significance in determining the future distribution of wealth and welfare in the world, has been their almost total neglect by the world press and other mass media. For the greater part these seem either not to have known, or not to have cared about, what was happening. This no doubt partly reflects the press's general disdain for everything that concerns the UN: a lack of interest said to be based on a belief that the UN is 'not news'. This is a somewhat circular argument, for the UN is 'not news' mainly because the press chooses not to make it news. Public *interest* would be greater if public *knowledge* was increased. In the particular case of the sea-bed, the press's unconcern is the more astonishing as the subject in itself is intrinsically interesting, and indeed in many ways fascinating, combining as it does much of the mystery and enchantment of the deep-sea region itself with the complexity of high diplomacy and the drama of international politics. There is little doubt that for these reasons it will eventually arouse the intense interest and concern of the general public all over the world and become the great international issue of this decade.

A second special feature of the sea-bed issue is the fact that it is, to an unusual degree, an international and universal one, affecting every nation of the world. It is not unique in this. There are a growing number of questions which are world issues in this sense: satellite communications, the world monetary system, international trade rules, the problem of hijacking, and the trade in narcotics might be given as examples of the increasing range of subjects which can be settled only at a world level. The one that is perhaps closest in character, and which has sometimes been compared to it, is that of outer space; another hitherto almost unknown region which has suddenly come within human reach, and for which international rules have had therefore to be devised. But though in character that issue is somewhat similar, in scale and importance it is wholly unlike: as outer space contains few resources that are likely to become available for human use on earth within the foreseeable future, all that was mainly required in that case was to ban military uses and attempts at national appropriation, and to devise a few other regulations concerning the notification of space-shots, the return of astronauts and so on. The sea-bed, on the other hand, not only contains valuable resources on a fabulous scale but will also very shortly begin to be exploited and available for human use. And as every nation felt that it had rights to a share in these resources, the question has inevitably been discussed with an intensity and passion that has never been accorded to the empty wastes of outer space.

The unusually international character of the issue has had a number of effects. It has meant that the question has been discussed from the beginning under the auspices of the UN. There have been no bilateral, or regional, attempts to resolve it (though there have been regional attempts to *influence* discussions, as in several Latin-American conferences). Nor have there been any *ad hoc* conferences, called specially for the purpose outside the UN, as occurred, for example, over the Antarctic in 1959. Because the issue affected all governments to some extent, there has been intense competition among members to be represented in the discussions; as we saw, the committee set up to consider the matter eventually contained almost two-thirds of the UN's members and almost every nation now attends the conferences. But the most important point is that it is generally accepted that the final outcome of the debates will be a universal treaty, to which all nations should become parties. Except in the case of the UN Charter itself, there has been virtually no other treaty among states in history for which universal adherence, or almost universal, was regarded from the beginning as essential.

This leads to the third unusual feature of these negotiations: the absolute necessity of securing consensus at every stage. In this it represents, in a sense, a reversion to the situation that existed in the nineteenth century and earlier, when universal agreement among the participants in a conference was always regarded as necessary to secure the passage of a treaty or other convention. In recent times, conferences and international discussions have been more often conducted in the expectation that at some stage there will be a direct trial of strength, and that decisions will be taken on the basis of a majority vote. In the case of the sea-bed, however, for the final treaty to be effective it is essential that virtually every major state becomes a party to it. If only two or three important states decide to go their own way, and seek then to exploit sea-bed resources outside the régime, this could fatally weaken the régime's effectiveness—even destroy it altogether. The fact that consensus is, as is generally admitted, so acutely needed, enormously complicates the entire bargaining process. For the need for consensus strengthens the bargaining position of every member. It means that not only every group, but to some extent every individual nation, has only to dig in its heels, and be really obstructive to prevent any agreement contrary to its interests coming about. This particularly favours the very large powers, since it is their non-participation which would be particularly damaging to the régime. But everybody, except the very small, shares the advantage to some extent. Where a large number of different groups, with radically conflicting interests, not to speak of many individual governments, all share this power, so that the most extreme view on one side has to be reconciled with the most extreme on the other, it is not surprising if negotiations sometimes become rather heavy weather.

A fourth feature which makes the issue unusual is the fact that it has been discussed until now in a single relatively enclosed international forum: the Sea Law Conference and its predecessor, the Sea-bed Committee. The initial exposition of the problem, the discussion of the various alternatives, the exposition of different view-points and, finally, the detailed negotiations have taken place exclusively within the same intimate body. There has been a little, a very little, desultory writing about the theme elsewhere. There have been occasional bilateral consultations between governments. But all the substantive exploration has taken place, from the beginning, within the small private club of the Sea-bed Committee's participants. This has met under the same chairman throughout. Many of the chief participants, the national representatives, have remained unchanged throughout. They have begun to acquire a considerable personal familiarity with one another and an understanding of the view-points of opposing governments. Although there are of course other issues which have been mainly discussed within a single UN agency or committee, none has been of the scope, complexity and importance of this one. Rarely have so few people been engaged in deciding the destiny of so many (and with the knowledge of so few) as in the discussions on the sea-bed.

A fifth feature has been the unusual character of the alignments which have been conjured up. In most multilateral diplomacy in the modern world there are certain relatively fixed groups among states which occur over and over again: rich against poor, west against east, conservatives against progressives. None of these traditional categories has had much relevance to the sea-bed discussions. In practice, every nation had an *individual* interest, according to its own geographical and technological situation. Where groups emerged at all, they were of an unfamiliar kind: coastal-developed, coastal-developing, land-locked, deep-seafishing groups and so on. There are, as we saw, at least nine or ten different interest groups of this kind. All the familiar blocks were divided: the developing among coastal and non-coastal, the developed among nationalist and internationalists, among oceanologically equipped and otherwise. Even the Communist block and the Latin Americans, usually the two most cohesive groups, were not entirely at one: this was itself a notable feature, especially in the former case (Mongolia and Rumania in particular, increasingly asserted their own interests in voting differently from the rest of the Communists; even Poland presented proposals significantly different from those of the Russians).

Nonetheless, the discussions retained some of the traditional features of UN diplomacy. There were, as in the UN, prolonged games of draft and counter-draft, amending and re-amending until eventually it was possible to arrive at a version that most people could live with. One of the remarkable features was indeed, given the importance and diversity of the issues, that it has been possible to arrive often at some degree of consensus. A good many of the resolutions in the Assembly have been passed by large or even overwhelming majorities. The Declaration of Principles, which was of considerable importance, was passed almost unanimously, with no votes against and only a handful of abstentions. And within the conference there has been substantial progress on the less contentious points.

Even over the basic issues there took place a slow convergence in many cases—a process characteristic of UN negotiations. There are a number of reasons why this occurs. In some cases there are straight deals; one nation or group drops one position in return for somebody else dropping another. In other cases, a nation slowly abandons a certain position simply because it finds itself in isolation and recognizes that it is pointless to maintain its stand with little support. Again, friendly nations may urge a government to drop some particular standpoint in order that they may stand more solidly together on others. Above all, there are strong pressures in favour of conformity: in every UN discussion, delegations feel acute discomfort at finding themselves isolated or in only a very small group, and actively seek and urge their governments to adopt positions that are nearer to the majority view.

Certainly such a process took place during the sea-bed discussions. In the early stages, a number of governments—including some of the most powerful—were against even the principle of a 'régime': they wanted only 'internationally agreed arrangements', or a set of rules for sea-bed exploitation. The Soviet Union and its allies were against the idea of setting up 'machinery', with the supranational overtones that this implied. Many more were against the common heritage idea. Even the idea that there could be no national appropriation of the area was not universally accepted in the beginning, but was everywhere adopted at a later stage. At first some governments rejected the idea that licensing was necessary, contending that a simple registration of activities would be sufficient, another standpoint almost universally abandoned later. And the proposal for the establishment of a sea-bed authority with wide-ranging powers, which at first seemed to many over-ambitious or even outlandish, was eventually universally accepted. These are only a few examples of points, once controversial, which eventually became through social pressures non-controversial, a part of the conventional wisdom.

The fact that this process took place does not of course prove that it will continue, and that an agreement will finally be concluded; it is inevitably the most crucial and controversial points that are left to the last. But it is perhaps a sign that there is at least a fair measure of goodwill available. Considering the importance of the issue, there has so far been a reasonable readiness to give and to take (though a few have so far shown themselves rather more adept at taking than giving).

Goodwill alone, however, may not be enough. Progress so far has been slow. And time is limited. Whether agreement is reached and in what form will depend not only on the conference's discussions but on how exploitation itself develops meanwhile. And it is to the future of exploitation in the area that we must next turn.

The Future of Sea-bed Production

The prospects for the exploitation of the sea-bed will affect both the form of a future régime and the speed with which it comes into effect.

Whatever the boundary finally chosen for the area, it is likely that the exploitation of surface nodules will take place before drilling for oil and gas. *Production* of oil and gas (as opposed to test-drilling) is today almost entirely at depths of 100 metres or less. It will probably be a number of years before much production takes place even at 200-metres depth, and still longer before it occurs in the international areas. Technically, this is possible already. It was hoped that production would be possible at 400 or 500 metres by 1980. Exploration is already going on at about that depth. But exploitation remains highly expensive in deeper waters, and the shallower deposits are certain to be exploited first.

Nodules, on the other hand, are thought to be most effectively exploited at depths of 4,000 or 5,000 metres (where they are most abundant and richest in metal). The equipment now being developed is designed to operate at such depths. As soon as such exploitation takes place at all, therefore, it will certainly be within the international area, and all the signs are that this will take place in the very near future.

A number of industrial corporations, mainly U.S.-owned, but with Japanese and West German participation, have undertaken vast investments in this field in the last few years. They have already conquered many of the most difficult problems.

First, extensive exploration of the nodules has been undertaken, especially in parts of the Pacific (for example, in an area close to Hawaii). Submersibles, using underwater photography and television, have been used to investigate in detail the deposits within these areas. These explorations have shown that nodules of 'exceptionally high content of copper, cobalt, nickel, manganese and other minerals', are present there, often at depths approaching

4,000 metres.[1] Discoveries have also been made of numbers of nodules buried *beneath* the ocean floor, but it is not yet certain whether these can be economically exploited.

Secondly, there have been very large expenditures on the development of the types of dredges required for recovering the nodules. A U.S. firm, after spending $16 million over a number of years, has successfully tested a hydraulic dredge system which has operated in 800 to 900 metres of water in the Blake Plateau in the Atlantic. The dredge works something like a vacuum cleaner by sucking up a stream of nodules, air and water, which are later separated. The same U.S. company is now building a larger-scale production model, capable of operating at 4,000 metres and of dredging about 5,000 tons of nodules a day. A Japanese company claims to have demonstrated successfully a continuous line bucket system, dredging at depths of 1,000 to 4,000 metres. This work is receiving extensive support from the Japanese government. A new international syndicate is now being organized by several large companies to develop a complete deep-ocean mining system, including ore carriers, supply ships and processing plant. The syndicate expects to start exploiting nodule deposits in the Pacific within a few years (as also does Howard Hughes).

Thirdly, progress has been made over the problem of processing the nodules to separate the metals. A good deal of work has been undertaken on this, not only by commercial undertakings but by the U.S. Bureau of Mines, by the University of California, and other bodies. Those working on the problem have concluded that 'extraction is economically feasible without radical or major departures in previous extractive metallurgical technology'.[2] Suitable methods are said already to have been discovered for undertaking the separation process at a reasonable cost. One major undertaking in this field, Deepsea Ventures, Inc., has announced the successful conclusion of tests of a pilot plant for processing one ton of nodules a day and has now begun manufacture of an operational plant. A UN Secretariat study stated: 'it now appears that the mining and metallurgical extraction of copper and nickel from the nodules is not only within the range of feasibility,

[1] *Mineral Resources of the Sea*, Report of the Secretary-General, April 26, 1971, E/4973.

[2] Ocean Science and Engineering Inc., 'The Economics of Offshore Mining', Paper prepared for the Resources and Transport Division of the UN, Jan., 1971.

without requiring major technological breakthroughs, but it is reasonable to expect such technology to be both economic and profitable within a few years'.[1]

Thus developments are taking place in a variety of fields which may make full-scale ocean production technically feasible within the next few years, certainly before 1980. It seems unlikely that so much expenditure would have been undertaken unless it had been concluded that such production would also be economically profitable. How far it may not merely be profitable, but could displace normal land production obviously depends on a number of factors. The main one of course will be the cost of the sea-bed operations in relation to those on land.

Estimates have been made of the likely economics of sea-bed operations. Assuming an undertaking under which 4,000 to 5,000 tons of nodules a day are collected at a depth of 5,000 metres, the cost of mining a ton of nodules, including capital costs, interest and insurance, has been estimated at various sums between $5 and $13 a ton.[2] Transportation costs to a processing plant on the coast—assuming a plant about 3,000 miles away from the dredging operation (for example on the west coast of the United States from a Pacific operation), and assuming two ore carriers of about 100,000 tons each—would be about $2.7 to $5.5 a ton, again including capital costs. Thus the total costs of the whole operation might be $25 to $35 a ton, plus perhaps $3.5 to $7 for depreciation. Using another system for extraction, total costs have been estimated as $40 to $50. The total capital cost for one operator, who might produce 1 million dry nodules a year, could be $200 million. Such a system could produce 280,000 tons of manganese, over 14,000 tons each of nickel and copper and nearly 3,000 tons of cobalt a year.[3]

These are of course only estimates, based on the best knowledge so far available. But the actual costs could as easily be lower as higher than these estimates. And all previous experience would lead one to assume that costs may well decline as operations extend. Enormous sums are now being spent on the development of the technology and production facilities. This is presumably based on a careful estimate of the likely costs, the market and the

[1] Report of the Secretary-General, *Mineral Resources of the Sea*, E/4973 of April 26, 1971.

[2] See UN Doc. A/AC138/36, pp. 48–9.

[3] *Ibid.*

eventual profits. Thus, by those best fitted to judge, sea-bed exploitation has been calculated as likely to be successful and profitable reasonably soon.

If sea-bed operations are competitive with land production in the near future, what will be the effect on existing producers? An exact comparison with the cost of land production is not easy. One of the main difficulties arises from the fact that production of the four main minerals concerned in nodule exploitation (manganese, copper, nickel and cobalt) must occur in the fixed proportions in which they occur in the nodules. This proportion is roughly 97 tons of manganese, 5 of nickel, 4.9 of copper to 1 of cobalt. But that proportion is entirely different from the proportion in which the four metals are sold in world markets. This latter proportion is 381:27:279:1. What this means, to take an extreme example, is that, if sea-bed production was directed mainly at copper and was able to satisfy 100 per cent of copper demand, it would at the same time satisfy 1,400 per cent of the demand for manganese, 1,000 per cent of the demand for nickel and 5,500 per cent of the world demand for cobalt (see Table 5). In other words, the markets for the three latter minerals would be totally swamped, the price would collapse, and all land production would be completely out of business.

Though this sounds somewhat fantastic, it is not an altogether impossible result over the long term. Eventually, nodule exploitation might be geared primarily to the production of copper, with the remaining metals regarded as side-products. However, given the size of the copper market, and the costs of sea-bed investments, it is perhaps more likely that, at least in the early years, sea-bed production will take up only a relatively small part of it. But even if sea-bed production met only 10 per cent of the world copper demand, the associated production of the other metals would still meet 140 per cent of world demand for manganese, 100 per cent of world demand for nickel and 550 per cent of world demand for cobalt: so virtually eliminating all land production of these minerals.

One way of estimating the proportion of each metal that is really likely to be taken by sea-bed production is by assuming a particular number of average-sized operators and estimating their production of each mineral. If one assumes that each sea-bed producer aimed at dredging 5,000 tons of nodules a day (even this is less than some systems that have been developed are aiming at),

Table 5

Simultaneous Availability of Four Associated Metals in Manganese Nodules for Alternative Hypothesis of Supplying the Entire 1968 World Production of Each Metal

Primary metal to be recovered from nodules	World production in 1968 Metric tons of metal content in ore	Assumed yield of metals per ton of dry nodules mined (kg)[1]	Percentage of 1968 world production of associated metals that would be made available simultaneously			
			Manganese	Nickel	Copper	Cobalt
Manganese	7,700,000	279 kg	25.4	18.4	1.8	100.0
Nickel	549,100	14.4 kg	100.0	72.4	7.1	393.5
Copper	5,473,000	14.1 kg	138.2	100.0	9.8	543.7
Cobalt	20,000	2.88 kg	1,406.4	1,017.9	100.0	5,534.1

Source: UN Doc. A/AC138/36.

[1] Based on the following assumptions:
(a) metal content in nodules: Mn 30 per cent; Ni 1.5 per cent; Cu 1.5 per cent; Co 0.3 per cent;
(b) recovery factor in metal separation: Mn 93 per cent; Ni 96 per cent; Cu 94 per cent; Co 96 per cent.

and that the recovery rate after processing was in the range of 95 per cent, then only five operators would be able to supply 0.64 per cent of the world demand for copper, 6.5 per cent of world nickel demand, 10 per cent of world manganese demand and 40 per cent of world cobalt needs as shown in Table 5. With only 10 enterprises of this size, almost all of the world cobalt market, a fifth of the world manganese market, and an eighth of world nickel needs, but little over 1 per cent of the world copper market would be met.

There are a number of paradoxes about the economics of production which result. At first sight, sea-bed production should be aimed primarily at the market for manganese, because manganese would be produced in much higher quantities than the other metals, and so, even given the slightly higher costs of processing, would still be far the cheapest to supply. However, against this, the *demand* for manganese is not likely to respond to reductions in prices, or to expand in relation to increased production facilities. Manganese is used primarily for a single purpose—purifying during steel production—and this market, though rising slowly, is not elastic. This means that, as sea-bed exploitation increased, the price would fall and profitability would disappear, both for land-based and sea-bed sources of production. Moreover, some land producers are low-cost and efficient, and might be able to cut their prices and remain profitable. It is likely, therefore, that sea-bed production of manganese would be attractive only for those countries in which the extraction process for the nodules actually took place: because of the savings in transport costs, and perhaps for balance of payments reasons, or on strategic grounds.

Again, one might have expected the production of cobalt from sea-bed sources to be attractive and profitable, because it is by far the most valuable of the four minerals. Though it occurs as only a small proportion of each nodule, it could be processed relatively easily; and sea-bed production would represent a substantial addition to a very small world market: only about 20,000 tons a year altogether. Here again prices might be depressed by the accession of this additional production. In this case, however, new uses might be found for cobalt, for example as a substitute for nickel in plating, special alloys and steel products, as its prices became lower. But because the total market for cobalt would still be small, it is doubtful if cobalt production alone could justify the very large investments to be made in sea-bed operations.

Thus cobalt, like manganese, is more likely to be a side-product than the main object of exploitation.

To justify its heavy costs, sea-bed production would probably require a market which is large in itself, but not inelastic like that for manganese. This suggests that the main markets aimed at will be those for copper and especially nickel, for which demand is rising rapidly. If ten operators of the size described entered production by 1980 (which might require $2 billion of capital), production from nodules would still equal only about 13 per cent of world nickel and 1.3 per cent of world copper demand, so it is less likely that prices of these two products would be seriously affected, at least in the case of copper. Sea-bed production might simply make up the increase in demand that would anyway be expected by that time (demand for copper is expected to increase by 4 to 5 per cent a year for the next 30 years and nickel by 6 to 7 per cent). On the other hand, the quantities of manganese and cobalt obtained as by-products from this would represent far larger proportions of world demand and, if produced and sold in addition, would severely depress the prices of these products. Even in the case of copper and nickel, sea-bed production would of course prevent the *growth* of revenues for the land producers that might otherwise have been expected. And it is likely that, in time, far more than ten operators would be engaged in sea-bed production.

The extent to which sea-bed production develops and displaces land production in practice will depend very much on the scale of the payments demanded by the international machinery (and by national governments, if licensing is to them in the first place). If royalties were made very low at first to encourage sea-bed development, this would in effect give a kind of subsidy to sea-bed production in relation to the land producers, who normally pay both taxes and royalties to one or more governments. If the royalty payments were made higher than the average of payments made on land, on the other hand, sea-bed production would be penalized, and so would technical advance. The logical course would thus clearly be to fix the total royalty payments for sea-bed production at roughly the average of the payments made by land producers.[1]

[1] It would also be logical for the payments to be based on the tonnage of production rather than on profits. In this way they would act as a fixed cost of production. Profits are always difficult to assess accurately, and in any case a levy on them might not secure to the authority such high revenues.

Even if one assumes royalties at this rate, it seems likely, on the basis of the estimates suggested earlier, that sea-bed production might soon be competitive with production on land. But in fact it is not certain than an equivalence in the level of royalties would be accepted, or accepted for long. As only a relatively small number of land producers would be affected, whereas very large numbers of the developing countries might gain eventually from increased production from the sea-bed, support could be given for lower royalties in the early years in order to encourage production, which might make sea-bed production still more competitive. So why kill the goose that could lay such golden eggs by unnecessary hardships? it will be argued. And especially if proposals had already been made for compensating land producers, by whatever means, it might be thought justifiable for sea-bed production increasingly to displace the traditional source of supply. But in any case the likelihood is of a rapid growth in sea-bed production of these minerals in the coming years.

So far as oil and gas are concerned, the situation is a little different. Whereas nodule exploitation will almost certainly *begin* in the international area, hydrocarbon exploitation will be only gradually edging out towards deeper and deeper waters over a period. Although for nodule exploitation some of the cheapest sources may be in deep waters (because this is where the nodules are most abundant and richest in metals), for oil and gas the cheapest exploitation will always be in shallower waters; the only possible exception is where very large reservoirs are easily tapped, perhaps stretching out from shallower waters. As there are at present alternative supplies of hydrocarbons, both on land and, especially, in the continental shelf, it may be some time before any substantial proportion of oil and gas supplies are obtained in the international area. But with the world energy gap, this may soon change. Certainly the total value of sea-bed hydrocarbon production will probably soon be greater than that for nodules. Thus, even if only a small proportion of total world supplies of oil and gas came from the area, it could nonetheless provide a large addition to the revenues of the international régime.

It is quite likely that gas may be produced before oil from the deep seas. This is, first, because of the astonishingly rapid increase that is taking place in the demand for natural gas: even faster than that for oil. Demand for natural gas is likely to triple in the next decade (against a doubling in the demand for oil). It is unlikely that this demand can be satisfied from new fields on land. On the contrary, it is the impending depletion of U.S. land sources of natural gas that may above all hasten the need for rapid exploitation of deep-sea sources. The United States at present produces about 60 per cent of all the natural gas produced in the world. Over and above that, it is also the world's largest importer of gas. But it is using up its own supplies so rapidly that at the present rate of consumption, its existing known reserves on land will be almost exhausted by 1985. This will create a critical short-fall to be made up at that time. Although part of this may be met from the Soviet Union, from new finds within the United States, Canada and Alaska, and from the importation of liquified gas, there will in all probability still be a very large gap remaining. This could make the exploitation of gas from the international area, where there are large supplies off the California coast, more attractive and profitable. The recent rapid rise in the price of oil and coal, and the current preoccupation with securing clean fuels to avoid pollution will increase the attractiveness of natural gas. Gas from the U.S. off-shore areas would also save the transport costs of imported oil or gas. And it could afford substantial balance of payments advantages, a matter of pressing concern to the United States in recent years.

Though oil production from the sea-bed may develop more slowly, in the long run it will no doubt be still more important than that of natural gas. Oil is already by far the largest single item in world trade, both in value and quantity. Within the continental shelves, oil production is worth three times as much as that of all other products combined, even including natural gas. And total production of petroleum from land and sea is worth three times as much as production of natural gas, copper, nickel, manganese, and cobalt combined. This gives some idea of the relative importance that oil is likely to have eventually within the economics of a sea-bed régime.

Some of the same factors which will promote the development of natural gas will also promote that of oil. About half of all oil production is at present concentrated in the United States and the Soviet Union. In the former case, at least, unless there are substantial new finds, land resources which can be economically exploited will begin to be exhausted within about a decade. Given the scale of production and consumption in the United

States, any such short-fall would be very large in absolute terms. At present both the United States and the Soviet Union are largely self-sufficient in oil. As the United States becomes a major importer, and with the rapidly increasing demands from West Europe, Japan, China and many developing countries, the total new supplies to be found will be very large indeed. There is no over-all shortage of oil in the world at present: the Persian Gulf alone has proved reserves amounting, on a very conservative estimate, to 60 times the 1970 production: and it is generally thought that a more realistic estimate would be well over 100 years' present production. There are also large supplies in many other parts of the world, and new finds being made all the time, especially in the continental shelf. But the huge rise in demand and price expected is certain to lead to the production of greater and greater proportions from the deeper parts of the ocean.

Technically, this could happen almost immediately. Production at depths of 500 metres will probably be possible by 1978, and at 1,500 metres within a decade. But at 450 metres production would be something like three times more expensive than at 100 metres and infinitely more expensive than production on land. Thus in general, exploitation of the shallower areas will precede that of the deep ocean floor.[1] But as we saw in Chapter 2, off-shore production in home waters may be preferred in some cases even despite a substantially higher cost. It may be more advantageous for balance of payments reasons; for strategic reasons; to save royalties to the producer government; and, above all, to save transport costs, which are an important part of the price of oil (a barrel of oil which only cost 10 cents to produce in the Persian Gulf, including return on capital, costs well over ten times as much to transport to Western Europe).[2] Expenditure on off-shore production is

[1] 'Given the high cost of deep-water development and the availability of petroleum beneath the continental shelves, petroleum production from the small oceanic basins, continental rises, and the outermost part of the continental shelves is likely to be restricted during the next one or two decades to a small number of very large highly productive fields in convenient locations. It implies that future petroleum development in ultra deep waters would involve systematic exploration of vast regions to seek out only the very largest oil fields that could be efficiently produced with a small number of highly productive wells which would justify the large investment in the necessary deep-sea production facilities'. *Mineral Resources of the Sea*, Report of the Secretary-General, E/4973 of April 26, 1971.

[2] $1.25 in the first half of 1971. Royalty and tax payments, mainly in his own country, multiply the price to the consumer many times further.

increasing at present by about 18 per cent a year and this is likely to continue. Especially when big reservoirs are found, or in geographically convenient situations, these may be regarded as worth exploiting: an obvious case will be in the Santa Barbara Channel off the California coast, where there are thought to be very extensive deposits at depths up to 400 metres.

The economic potential of sub-sea oil and gas from beyond the shelf will depend on the levies imposed by the international machinery. Here too the logical course would be to impose levies which have the effect of putting deep-sea production on a roughly equivalent footing to that from land or shelf sources: given the scale of production that may eventually emerge and the level of these royalties, this should provide eventually a substantial revenue for the international machinery—far greater than it would obtain from the exploitation of nodules alone. Neither oil nor gas from these sources is likely to be disruptive of prices. For in this case, sea-bed exploitation will not bring a sudden access of large-scale new supplies on to the market. What will occur will be a relatively gradual growth of production from deeper waters, under conditions not unlike those for existing production. Because demand itself will certainly continue to rise sharply, for both oil and gas, prices are almost certain to continue to rise. One at least of the problems affecting nodule exploitation may thus here be less acute: the effect in disrupting the market and displacing existing sources of production. But the greater scale of the revenues obtained from oil and gas may, on the other hand, bring still more intense controversy over the system to be employed, and the amount of competition with supplies from national areas which should be allowed. It presupposes too a fairly narrow limit.

These projections of the future production of nodules and hydrocarbons are highly speculative. New developments of various kinds—for example, new discoveries on land, a deterioration in the expected economics of sea-bed production and other imponderables—could throw the calculation awry. But these could scarcely alter the fundamental problem. They would merely change the time-scale, and that relatively marginally. Whatever happens, there seems today to be an overwhelming probability that by the end of this decade and perhaps substantially earlier, there will be a significant volume of mineral production from areas lying beyond the continental shelf as traditionally understood, and certainly well beyond the geological shelf.

The next question we must therefore consider is: when that happens will an international régime have been established and be able to cope with the problems that arise?

The Future of the Régime

To devise a method of government for two-thirds of the world's surface is no light task. Negotiation on *specific* issues, as against the general exchange of views on types of system, is only now beginning to take place. This process will be a long, stormy one. Not only are the interests to be reconciled among different nations and groups complex and conflicting. The whole range of issues is vast. On some lesser questions the process of convergence, the whittling away of differences through hard bargaining, the exchange of one concession against another, leading finally to some measure of consensus, will no doubt continue, slowly but nonetheless surely. But on some of the major issues—the voting system within the authority, the criteria to be used in licensing, whether there should be a limit on sites for one nations' companies—on these and many others, there could come about a prolonged deadlock, with a variety of groups holding obstinately to their different views. These groups are unlikely to be willing to give way on the basic issues at all easily. And it is quite possible that, even when the last conference has been held and concluded, no final agreement will have been reached.

There are two divisions of interest which outweigh in importance all others: that between coastal and non-coastal states; and that between developed or, rather, technologically equipped states (by no means the same thing) whose nationals might be expected to undertake the actual exploitation work and the rest, the great majority, being consumers of technology. But in neither case are these groups homogeneous; there are wide degrees of 'coastality': a country like Brazil, with a very long coastline, has an interest altogether different in degree from one like Costa Rica, say, whose coastal shelf area is certain to be extremely limited whatever boundary is used. Similarly, among the technologically equipped states, the interests of the United States, whose nationals and companies comprise perhaps over three-quarters of those likely to be directly involved in exploitation, and of a country such as, say, Sweden or the Netherlands, which are unlikely to play a major role in exploitation, whatever the system used, are totally different.

It is possible to envisage a situation, perhaps in two or three years' time, when a number of the less contentious issues will have been virtually disposed of. These would include both those questions that did not matter too terribly to anybody; and those in which the countries which did have strong views were a relatively small minority and were eventually ready to make concessions. In the former category might be questions concerning the structure of the authority (except for voting rights), the subsidiary commissions and bodies which might be involved, the technical regulations governing exploitation, the size of the blocks, the commitments regarding protection of other uses and the avoidance of pollution. None of these is in a fundamental way contentious. Among the latter might be the concern of the Communist states to avoid any supranational authority; as they are in this respect a very small minority, they may find that they have to give ground (as they already have done to some extent) if their views are to have any weight at all. Similarly, the countries, such as Britain, demanding total freedon for their fishermen to fish anywhere outside territorial waters will find themselves increasingly isolated, and may be more prepared to consider a compromise providing for preferential and other fishing rights beyond the twelve-mile line, and so on.

But there will remain a number of issues on which the major interest groups will find themselves still totally at loggerheads. On the question of the limits—the question which above all others can tilt the balance of advantage between the coastal states and the non-coastal—there is unlikely to be agreement before all else has been agreed. Even then there could be a situation of almost total impasse, with two roughly equal groups refusing to give way: most of the coastal states on the one hand demanding wide limits, with the non-coastal, the small coastal states and a few internationalists on the other, calling for narrow limits. On the questions which divide developed and developing too—voting rights, the level of royalties, the system of distribution, whether there should be a time-limit for the initial system, the discretion of the international authority and the sanctions available to it—it is equally difficult to see agreement being reached easily. Here the balance of numbers will be quite different. Not only will the developing countries wield nearly three-quarters of the votes, but they would also have the support, on some issues at least, of a few developed states, especially in West Europe—Sweden, the Netherlands, Belgium, Denmark and others—which have shown every sign of supporting the concept of a fully international régime, with the maximum area and the maximum revenues for it. But they will have against them a few of the most powerful states in the world.

The group of entrenched nationalists holding out for the minimum of international control, for low royalties, and for heavily weighted voting in favour of the larger and richer countries, will form a relatively small minority. They will, however, include some of the world's most powerful states: two of the three superpowers (China's eventual policy on the whole question still remains something of an enigma),[1] and several of the largest of the remaining powers, including Japan, France, Britain, Italy and probably West Germany. At present the views of this group are by no means united.[2] It is possible that in view of the confrontation in which these countries will increasingly become involved against the poor countries, they may soon begin to co-ordinate their positions to appear more at one (the Soviet Union has already sought the support of Britain and other countries for some of its proposals, while Britain has moved towards support for the U.S. revenue-sharing concept, which it first rejected).

With little doubt this will be one of the major confrontations. Perhaps four-fifths of the nations of the world will be ranged against a small minority of rich countries, including the United States and the Soviet Union. The former will stand for a fully controlled system, high royalties and, in some cases, international operation; the latter for a much freer system, with low royalties and little international working. It is a confrontation that will probably be bitter and protracted, and it may be worth sketching out various scenarios of the situation as it could develop in later years.

It must be recalled, first, that while these discussions are taking place—that is, during the next two or three years—the situation as regards exploitation will not be standing still. We have seen the

[1] China has supported demands for a 200-mile limit, which would correspond with its interests as a large coastal country, as well as its political interest as a supporter of developing countries against the United States and the Soviet Union.

[2] The United States believes in a relatively strong régime, and a narrow national limit for many purposes, but a wide limit for national control of exploitation; the Soviet Union believes in a weak system for all purposes, with the minimum of international régime and low royalties. Britain and France believe in a wide margin, and a considerable degree of national control even within the international area.

speed with which development has already occurred in the last few years. As recently as 1967, when Dr Pardo first sketched out the possible dangers (and was thought by many to be exaggerating), nodule development seemed still to be far in the future and there was little practical experience of it; already today equipment for dredging in several thousand metres and for the separation of the minerals has been developed and will shortly be in regular production. At that time the maximum depth for oil production was around 100 metres; today it is possible at around 200 metres, and is expected to be possible at 450 metres in the next few years. Already today active preparations are taking place for production in areas well outside the national area as traditionally understood. The companies involved in such preparations may not believe that an international system will ever be established; or they may believe that it will not affect production which has already begun; or that under the régime they will still operate even if under somewhat different conditions. The main point is that, by the time a régime is established, active exploitation may already have begun in areas which many countries believe should be part of the international zone.

What this means is that interests in arriving at a rapid settlement of the issues will begin to diverge radically. Until now, the rich countries have wanted a fairly quick decision on some points at least—for example, the breadth of territorial waters and fishing rights—quite as much as the poor. But in the conditions just described, the position would be that, whereas it became increasingly urgent for the poor countries to secure agreement as quickly as possible, it would be increasingly in the interests of the richer countries to delay and procrastinate as much as possible. The further exploitation had developed in the areas beyond the traditional shelf, such countries might feel, the stronger their bargaining position would become, and the less onerous the conditions they might be able to negotiate, for the international system.[1] This would in turn promote increasing impatience among other countries, and demands by them for the treaty to be agreed as soon as possible. Increasingly, the conference might be used not as a means of securing general international agreement but as a way of forcing through decisions designed to forestall private

[1] The United States administration has so far however firmly resisted proposals put forward in the U.S. Congress to allow U.S. companies to start operations unilaterally in the deep sea area.

development in the area, and to impose majority views concerning the régime to be established.

For example, there could be new attempts to impose a moratorium on private development. The Declaration of Principles (which forbade activities in the area except in accordance with the régime) will be widely quoted. If any company or government were none the less to claim that until a régime has been agreed, such injunctions have no effect, the response is likely to be a demand by the poor that the régime itself should be agreed immediately.

But such pressures are unlikely to succeed. For the effect of the general demand for consensus is that a small group of rich countries could hold up agreement almost as long as it liked if it had a mind to do so. Where there is no real desire to agree, agreement can always be postponed. And the poorer countries might find themselves faced with the alternative of capitulating to the demands of the wealthier states, or going without an agreed régime at all. Under these circumstances, the confrontation would become even more intense and bitter, with increasing impatience on one side and increasing resentment on the other. And meanwhile exploitation would gradually gather way. The chances of an effective régime would diminish all the time. A scramble for the most valuable deposits could begin: a new Klondyke in which those already richest would grab all the gold.

This is probably the most likely scenario. But so far we have assumed a straight rich–poor confrontation, with relative solidarity among the rich. There is, in fact, a big difference between U.S. and Soviet views, and a still bigger one between the views of coastal and non-coastal states. And alternative scenarios can easily be envisaged. The United States might seek to modify its position so as to appeal more to its potential allies, the other coastal states: a majority of those represented. It could do this relatively easily. It could decide to give support to a relatively wide economic zone even to the edge of the margin in return for a system allowing the widest possible opportunities for private companies and little discretion for the authority. The United States could thus hope to win support among many coastal states everywhere. The position would then be, even more explicitly than today, a straight battle between the coastal and non-coastal states, the most fundamental division of interest. The main points at issue would then concern the exact boundary where the economic zone, under this definition, ended, and the proportion of royalties going to national and international purposes within that area.

But the land-locked and shelf-locked states too might shift their position in order to win greater support. At first sight, the land-locked have a prime interest in a very strong régime with the maximum redistribution. But in fact they have two interests: one in a strong régime; the other (and much more important) in narrow limits. As a group, they could decide that to win the votes they need, they should seek to mobilize the support of the nationalists, especially perhaps of the Communist block. Three of the Communist block are themselves land-locked (Czechoslovakia, Hungary and Mongolia), and others (Bulgaria and Rumania) are almost so; even the Soviet Union does not have a strong coastal interest. Conversely, the land-locked themselves could be said to have some interest in a nationalist solution—that is, licensing to governments rather than to companies, as this could give them, for the first time, a direct interest in the sea and marine resources. This group might therefore shift its support towards something like the former British and French proposals—giving every nation a direct share in a part of the sea-bed, together with a relatively weak régime—provided that the limits were (unlike in the British and French plans) relatively narrow. This could be a means by which they might mobilize support against the uninhibited appetites of the coastal states as a whole—for example, among coastal developing states with small coastlines.

Alignments might change too on the question whether licensing should be to states or to companies. At first, a large number of delegates of all kinds assumed that licensing must be to states. This perhaps appealed to many developing countries as a means of giving them some direct interest and control over sea-bed exploitation in the future. As they begin to calculate their interests more accurately, an increasing number may favour direct licensing, on the grounds set out earlier in this book (pages 224 to 229). But others may continue to be attracted by the idea of controlling their own particular areas of the sea bottom, if only to acquire the opportunity of becoming, at last, colonial powers themselves. The developed equally could take divergent views on this. The internationalists, such as Sweden, Denmark, the Netherlands and others, may increasingly move towards supporting direct licensing as being ultimately a more internationalist system. So could the land-locked, which would anyway have the most difficult problems in administering their own areas. Other developed countries, such as the United States, France and Britain, will probably continue to favour licensing to governments, as their own plans assume, if only because this would reduce the power of the régime. The Communist states too, if they find that a system of free exploitation or weak control is impossible, could veer more in favour of the British and French concept of licensing to governments as the next best thing. So here too quite a different alignment from those traditionally adopted could arise.

Yet another break in the straight rich–poor division could occur. So far, most western governments have favoured a system of licensing to states. But it is not inconceivable that, if there were a very large-scale movement in favour of direct licensing, they might move in this direction in return for concessions elsewhere, such as a wide margin. This would be on the ground that their companies in any case could expect to win a large proportion of the contracts: from the national point of view, it was more important to win agreement to a wide national area than to increase direct revenues for their governments. The Soviet Union, however, is certain to continue to favour a system of licensing to states. For her, a system of licensing to companies will always be suspect. This is not because, as the Soviet Union has itself sometimes argued, a system of direct licensing is to adopt a capitalist system of exploitation: it would be equally possible for Soviet state enterprises to tender for contracts as for U.S. private companies. It is because it fears that the technical capacity of its enterprises does not yet match those of the United States and other western companies. At the same time, it may have suspicions that a licensing system controlled by others could be used in some way to prevent Soviet enterprises from playing a large role in the area. Above all, it would certainly wish to resist any proposal that might have the effect of strengthening the international machinery and the revenues of the international régime.

So far we have been considering possible developments that could take place in the near future. It is necessary now to look a little further ahead to what might happen later.

The conference will no doubt be spread over several years. Within two or three years, it is not unreasonable to suppose, there may be widely agreed draft articles on certain points, and perhaps whole draft conventions on others—for example on the high seas, on the contiguous zone, on pollution and preservation of the marine environment, even on fishing rights. But there will no doubt also be a series of major points on which it had proved impossible to reach agreement. The wide scope of the whole question, the large number of subjects which are in some way interrelated, the vital importance of the issues to many countries, sometimes for quite different reasons, all of these make it unlikely that there will be any quick agreement.

On the major issues—the position of the boundary, the nature and powers of the international authority, the scale of royalties, the system for the distribution of profits and so on—there might be no agreement at all. Indeed, on any rational calculation this is the position most likely to be reached by, say, 1976 or 1977.

Let us, however, for the sake of argument, make a more hopeful assumption and assume that on some of these major points at least—say, the powers of the authority, the system of distribution and even the voting system—a substantial measure of agreement had by then been reached. Only two or three issues of the most fundamental importance would perhaps still divide the conference: let us say the division of the boundary, the scale of the royalties, and the entities to which licences will be granted. If those issues were simple and clear-cut and closely related, it might just be possible even then to envisage that, finally, some great mammoth package could be negotiated, in which concessions on one point were met by concessions on another. Given the nature of the issues, however, it is difficult to conceive of this happening. For they in fact divide quite different groups of countries. The first divides coastal and non-coastal, the second rich and poor, and the last nationalist and internationalist. There is no conceivable trade-off that could easily be made among them.

Thus eventually there would be strong pressures to reach agreement by a majority vote on each issue. It is of course possible that those in a majority on each—say the coastal states on the first, the developing countries on the second and the nationalists on the third—were so far-sighted and generous that, despite their voting power, they were prepared to accept a compromise that gave as much to their opponents as to themselves. It is possible. But there is nothing in the history of international negotiations, or the attitude of the states concerned in this issue, which would suggest that it is in any way probable. Far more likely is it that, if the conference faced a final breakdown, on each issue the majority would seek to force through a vote on the lines they favoured themselves:

that is for a wide boundary, high royalties and licensing to governments alone. A paradox of this situation would be that though a majority had decided each issue, a majority would not be satisfied. The final result might satisfy almost nobody; only those countries which were at once coastal, developing and nationalist (say, Brazil, Chile and China). Very many other countries would have been defeated on one or other of the issues (and of course, there might be five or seven such issues to be settled in this way rather than three). And very many might therefore decide that, because they had been defeated on major issues, they were unable to ratify the treaty as finally agreed. And it would need only a fairly small number of nations to reach such a decision to make the entire régime unworkable.[1]

This is merely an indication, based on a highly simplified model, of the difficulty there will be in securing a final consensus on an issue on which, above all others, consensus is essential. Indeed the model is in some ways an optimistic one. It is perhaps more likely that the conference will simply fail to decide any of the basic points: either because there exists no two-thirds majority, or because of the objections of vitally important minorities which must be heeded. Or the conference could end with a straightforward confrontation on a single issue: perhaps between coastal and non-coastal states on the question of the boundary. A vote might be taken which gave to neither a two-thirds majority, so that the question remained undecided. This could alone prevent the régime from coming into existence at all. And in all these cases exploitation would then begin on a free-for-all basis.

Equally probable is a breakdown of another sort: through a straight confrontation between rich and poor, for example on the scale of the royalties, the form of distribution, the voting system in the authority, the obligations for training or other issues. Here a two-thirds majority could undoubtedly be secured by the developing for their views. But the result would be almost the same. Even if the draft treaty were approved, the régime would have

[1] This difficulty would be reduced by adopting the U.S. proposal that a provisional régime might be introduced once it had been adopted by the necessary majority, without waiting for ratifications. But it is not clear that this is intended to apply to the final treaty. The provisional régime would probably represent only an undertaking to pay a proportion of royalties for international purposes: since no authority for licensing would yet exist, the choice of site would apparently remain with the individual enterprise.

little meaning or reality if the major developed countries, including the United States and the Soviet Union, refused to enter into it. The mathematics of the voting figures would be irrelevant in the face of the underlying reality: the hostility of the major powers of the world, possessing the greatest capacity for practical exploitation and the power to enforce their views, could kill any régime.

This leads us to the third kind of scenario to be considered: the situations that could arise *after* a treaty has been signed and a régime established. Even if we assumed that a treaty of some sort does receive the two-thirds majority required at a conference, is subsequently ratified by the same number or an even larger number of governments, and that the régime consequently enters into effect, we must accept at least the possibility that some important governments may decline to sign or to ratify the treaty, and subsequently take no part at all in the régime. Let us assume, for the sake of argument, that the number of countries declining to participate is relatively small, say only 10 to 15 out of 145, but that it includes (as it very well might if the majority use their voting power to force through the régime they want) the Communist countries and the United States. How would events develop then?

In these circumstances the international sea-bed authority itself, and all those governments supporting it, would undoubtedly claim that exploitation could take place in the international area only under its authority and on terms which the treaty had laid down. The Declaration of Principles, which was supported almost unanimously, declared that 'all activities regarding the exploration and exploitation of the resources of the area . . . should be governed by the international régime to be established'. All the members of the régime would therefore hold that, even if they did not sign the treaty, the United States and Soviet Union and their enterprises could in these circumstances undertake no activity at all within the international area. They could only do so if they subsequently became parties to the treaty under its terms. The United States and the Soviet Union, however, might argue that, as they had declined to enter into the treaty establishing the régime, they were in no way bound by its provisions. No country could be subject to the terms of a treaty it had not explicitly accepted. And though they had previously accepted that all activities should be governed by the 'régime to be established' this had meant a régime which was generally accepted; that is, one agreed by themselves. They might even hold that any other régime was illegal, in that it purported to usurp the sea-bed for its own members without having secured the agreement of important nations that had a vital interest in the area. For all these reasons they would hold that they were free to undertake exploitation outside the provisions of the régime, on the same basis as if it had never been set up at all. And they would claim that, though international law was somewhat unclear in this respect, exploitation within the sea-bed would certainly be legal, either on the basis of the freedom of the high seas, or through the application of the 1958 Geneva Convention, or on some other plausible basis under existing international law. Many such arguments have already been found by ingenious international lawyers to support this view-point, and we quoted earlier well-known textbooks of international law which appear to acknowledge rights of exploitation throughout the area.[1]

Such arguments are by no means conclusive. It could plausibly be held that there existed no international law relating to the exploitation of the sea-bed beyond national jurisdiction before the treaty had been formulated: the International Law Commission had explicitly concluded that it would be necessary to create a new law in this respect in the future during its consideration of the law of the continental shelf. Under these circumstances, it could be maintained, there was no right to exploit these areas under existing law. Only the treaty itself would institute a form of customary law, which would be binding on non-members as well as members. Article 38 of the Law of Treaties Convention states that 'nothing in Articles 34 to 37 precludes a rule set forth in a treaty from becoming binding upon a third state as a customary rule of international law recognized as such'. Recent treaties which have been regarded as customary rules of law of this kind include the Conventions on the High Seas and that on Outer Space. A treaty on the sea-bed could reasonably be regarded

[1] The possibility that nations which had rejected the régime would still be free to undertake exploitation was accepted in a Secretariat study of the legal position, published in June, 1969. This concluded that 'in the event that a state declined to become a party to a treaty establishing international machinery relating to the exploration and exploitation of sea-bed resources, it would not acquire any obligation or rights under the instrument. . . . Assuming that the state in question did not accept any rights or obligations under the treaty, its activities would be based on existing customary and conventional law. . . .' (A/AC138/12/Add. 1 of June 30, 1969.)

as coming within this category. The special rapporteur on the Law of Treaties Convention, Sir Humphrey Waldock, commenting on Article 38, said that treaties of this kind could become regarded as general rules of international law 'either through the number of accessions or through general acceptance and custom'.[1] Both of these conditions could be claimed to have been fulfilled in the case of a sea-bed treaty accepted by the great majority of states. It could be argued too that a régime established for sea-bed exploitation had become an 'objective régime' in the sense used by international law: that is, a set of rights and obligations that are valid for all countries irrespective of any formal act of consent: treaties of neutralization, or demilitarization, or free navigation, and the UN Charter are sometimes given as examples of such objective régimes.

However, the legal arguments available are unlikely to be decisive. As so often in international law, equally eminent legal authorities will no doubt be found to support diametrically opposite views. The important fact is simply that those countries hostile to the régime established may deny that they are under any obligation to respect that régime, and proceed to act and to allow their nationals and enterprises to act as if they were totally free agents. They might (at least in the case of the United States) decide voluntarily to make certain payments, at a level which they chose for themselves, to the international community. But they would accept no other obligations than those which they undertook voluntarily in this way. More important, they would possess, and would use, the power necessary to defend this right. They might dispatch their naval forces to protect any enterprises of their own nationals which were active within the international area and defy any authority set up to prevent them from doing so. The countries which were parties to the régime might, conversely, use their own naval forces on its behalf. A major military confrontation throughout the waters and sea-bed of the oceans might be threatened: with China and Brazil, say, leading the poor countries against the United States, the Soviet Union and the rich.

However horrific this picture may appear, it is not altogether unimaginable. In the circumstances described—that is, if the United States and the Soviet Union were at one—the weight of

[1] Yearbook of the International Law Commission, 1964, Vol. II, 'Report on the Law of Treaties', Article 63.

power at least in the earlier years would be on the side of the developed powers. But it is almost equally plausible to envisage a situation in which these two powers had become divided. Although both rejected the régime, they might proceed to compete with each other for control of parts of the sea-bed, for economic or military reasons, each using their own navies to defend their own operations.

Alternatively, a system might be negotiated which was acceptable to one super-power but not to the other; in this case the navy of one might be deployed to enforce the régime, and that of the other to resist it. Chinese military power and that of other developing countries might be engaged on one side or the other. Given that a large part of the world's resources were at stake, protracted conflict might be waged all over the ocean-bed over many years. And a whole new range of undersea naval weapons, of the kind briefly hinted at in Chapter 3, might be called into action.

Gruesome though such a scenario may seem, it is by no means beyond the bounds of plausibility. The fact that it is even possible could well be used, especially by the two super-powers, as a bargaining weapon to warn developing countries not to risk overplaying their hands. It could be used as an argument in favour of adopting a type of régime acceptable only to the super-powers. And, increasingly, overt threats might come to be used as one of the principal weapons of negotiations in the discussions which are now proceeding.

Does this mean that, especially if they act in co-operation with each other, the two super-powers can exert a stranglehold over the course of the negotiations, and ensure that no régime is adopted which does not correspond with their own views and interests? The situation is perhaps not quite as desperate as this. For though the bargaining power of the majority may be weak from a purely military point of view, in other respects it is not inconsiderable. There is, for example, little value in exploiting the resources of the sea-bed if they cannot thereafter be sold. And it is not impossible that a universal boycott might be established against all minerals extracted from the sea-bed outside the provisions of the régime; or even a more general system of sanctions. The threat of such a boycott might at least act as a deterrent to exploitation which had not been authorized by the régime. The experience of Rhodesian sanctions, which for all their defects clearly induced some willingness to think again among the Rhodesian leaders, suggests that the threat of a boycott or sanction is not always altogether without effect. The general possibility of conflict and instability concerning the régime might be a powerful deterrent against unauthorized production.[1]

There are many other possible post-treaty scenarios. The non-accepting powers might be smaller in size but more numerous: say, some or all of the Latin-American countries, perhaps because of a decision on the boundary. These might seek to enforce some kind of regional system within their own area in defiance of the agreed régime. In this situation, some of the great powers, possibly at the instigation of the land-locked countries (which, whatever assurances might be made, would be the great losers from any regional system[2]), might seek to use force together to break up such a regional schism. Alternatively, the great powers might merely shrug their shoulders at any such attempt, either because they did not feel it worth the military effort and political odium to resolve the question by force or because they were otherwise content with the régime, or even because they were not unhappy that the entire régime was to be exploded. Or again, such a breakaway might encourage other coastal states or regions to go it alone in imitation, in order to secure their own interests. In this case, the entire international system might collapse, leaving only the land-locked and shelf-locked totally without benefit, but with different regions busily co-operating to develop their own areas. Perhaps the most likely outcome of all is that a general state of confusion of this kind could lead to a situation of grab-as-grab-can, on a 'first come, first served' basis, with the most technically advanced states taking all—precisely the situation a régime was intended to prevent.

There are still other possibilities. There could be bitter disputes, after the acceptance of a treaty, over the allocation of sites or the

[1] Many states have demanded that claims or operations contrary to the Treaty 'would not be recognized'. This could mean that finds being worked before the treaty came into effect were allocated to other operators, leading to considerable conflict and insecurity of investment. It is possibly on these grounds that the United States has now proposed the introduction of an interim régime, pending final agreement on a definitive treaty.

[2] Coastal developing states have proposed that each *region* should set up special arrangements to help its own land-locked states. But this represents a fragile assurance for the latter. Moreover the best technology might well not make itself available to developing *regions*, on the grounds of the high political risks involved. A world-wide system thus suits the land-locked better.

distribution of benefits, leading to walkouts from the system and unilateral action. Or again, a total breakdown could lead to deadlock and the establishment of a series of rival regional systems, which came increasingly into conflict over boundaries between them or other questions. Or some countries might join together to defend regional systems within their own areas whereas others sought to maintain a free-for-all system in the deeper waters, particularly for the nodules, so that totally different systems emerged for hydrocarbons on the one hand, and for nodules on the other. In any of these cases, the opportunity for a new international system would be destroyed. There would be no genuine international redistribution. There would probably be no system of training for developing countries. There would be no benefits to non-coastal states. The chance of embarking on an entirely new system of international redistribution would have been abandoned.

None of these is altogether an unlikely possibility. They of course only emphasize once more the vital importance of achieving consensus in the vital discussions to come. But this does not necessarily make it any more likely that such a consensus will be achieved.

The Challenge

There is thus every likelihood that the technology of exploitation will develop rapidly over the next few years. It is less certain that political skills and attitudes will develop correspondingly to bring about an effective and mutually acceptable régime to control exploitation.

Given the supreme importance of this issue, the size of the resources involved, and the sharply conflicting character of national interests, it is perhaps not altogether surprising if progress has been slow and disputes bitter. It is arguable that in this issue, mankind faces one of the greatest challenges of its history: possibly the greatest economic issue it has ever faced until this time. For on the decisions reached will depend the distribution of benefits from a large part of the resources of the earth. Perhaps more important still, on them will depend whether or not a wholly new era of human history is inaugurated, an era in which for the first time the principle of the common ownership of resources and their common exploitation for the benefit of all, which has been applied on a greater or lesser scale within nations for hundreds of years, is now applied for the first time within the world as a whole.

Ever since political society began, men have been arguing about essentially similar problems. What is the nature of the rights men have or claim in the natural resources around them? Should such resources be exploited for private purposes, or for the benefit of all? Are these resources to be regarded as common *spoils*, equally available to all on the basis of their competitive strength or skills, and their profits appropriated by those who can, and transmitted to their descendants, personal or national, after them? Or are they on the contrary to be considered as common *property*, to be shared equally by all, commonly exploited by all, for the common benefit of all?

Thus the issue itself is not new. What is wholly new is the context in which it now occurs. For here for the first time the resources at issue are not national resources. The issue is not: should resources be used privately, or shared among the nation as a whole? They are *world* resources. And the issue is: should they be used nationally for national advantage or shared among the world as a whole for the benefit of all peoples?

Political philosophers have always been the prisoners of the political society which gave them birth. So Greeks discussed and debated the form of the ideal city state, mediaeval thinkers the ideal sovereign, seventeenth- and eighteenth-century thinkers such as Locke and Rousseau, the ideal democratic state, early nineteenth-century thinkers such as Hegel and Fichte the ideal nation-state, Marx the ideal proletarian and socialist state. All talked about the state; but the idea of the state was modelled on the society in which they existed—in recent centuries always the nation. The idea of socialism in particular, born and developed in the era in which the nation-state was dominant, has above all always been conditioned, deformed and debased by the framework of that system. It has always meant the sharing of resources and wealth—but only among those who happened to be born under the same flag. Even its terminology has accepted this narrow objective. It is conceived in terms of establishing a 'socialist *state*'. Its major instrument has been thought of as '*nation*alisation'. 'Socialism in one country' has not been thought of as a contradiction in terms.

But socialism in this narrow meaning, always an inadequate concept, becomes today increasingly one that is totally irrelevant.

For if the word has any single core of meaning, socialism is about equality: about the sharing of wealth and activity to establish a more just and equal society. And today no amount of sharing of wealth and activity *within* nations can have the smallest impact on the major inequalities of the modern world. For those are not within states, but between them. Any form of socialism that is to be of the smallest value in confronting those inequalities can only be a socialism that is applied within the world as a whole.

The forces creating this inequality are of many kinds. There is the powerful gravity pull of the most advanced industrial areas in attracting still greater investment and development, as it is there that exist the high incomes, the large markets, the developed commercial networks and sources of supply. It is there that are the skilled labour forces. It is there that are the good transport facilities, good supply and servicing networks, the stable and efficient administrations. It is there that is the overwhelming concentration of research and development work and new innovation. It is there that established manufacturing facilities and labour forces encourage industry to build on assets already developed, rather than risk new ventures in the economically and politically precarious environment of strange countries in remote parts of the world.

Above all these factors, there is the process of collective accumulation. For today it is no longer the accumulation of individuals that serves to concentrate wealth and accentuate inequality. Saving is today primarily collective. On the one hand, there is the enormous corporate saving of large companies. This has the effect that huge profits are ploughed back into the enterprises which acquired them, or at least to the community where the shareholders are situated, rather than as once, dispersed in remote parts of the world where the risks are thought to be higher. On the other hand, national governments collect vast resources through taxation, and invest them in facilities, services and infrastructure in the states where wealth is already highest. Increasingly today it is the nation-state itself which, through taxation and public expenditure, hoards and re-invests the wealth of the community, building up enormous capital assets in the form of roads, railways, educational and health facilities, public amenities, research facilities, state industries and many others. Financial, trade and tariff policies are used to buttress and extend these advantages. Through both processes the states which already possess the greatest assets build up still greater ones, and further increase their capacity to acquire wealth in the future. The attractive power of the wealthiest states is still further increased; and they are able to draw to themselves not only investment, from within and without, but often the ablest and most enterprising among the labour force of poorer states as well, who are thus drained away and further deplete the exiguous national assets of their own countries. So, among nations as among persons, to them that hath still more is given.

The newly discovered resources of the sea-bed provide, for the first time in man's history, the opportunity to put into practice a form of world socialism. Here for the first time, the common ownership of common resources, and their use in the interest of all, may be used to bring about redistribution, however inadequate, not from rich individuals to poor within single states, but from rich individuals to poor within the world as a whole. The organized charity of national aid programmes, the welfare-state of the UN agencies, or the concealed domination of foreign investment programmes, could here be replaced for the first time by a genuine international socialism. And since a stage in world history has been reached in which even the poorest members of rich states are richer—far richer—than almost every inhabitant of poor states, it is a process of this kind which can alone redress the inequalities which are of primary importance in the modern world.

But there are other reasons, almost equally important, that make it essential that an adequate solution to the problem of the sea-bed is reached in the years immediately ahead. What has occurred over recent years is that man has acquired the capacity to occupy, to use and to exploit that vast region of the world, covering two-thirds of its surface, that lies beneath the seas. This has brought the urgent need of a system for governing that region. If it is allowed to succumb to a free-for-all, to be pillaged, polluted and annexed, to become the booty of an unrestrained competitive struggle, as in some of the scenarios just described, disasters of many kinds become almost inevitable. Environmental damage of the gravest kind may be caused to the area. Military dangers of almost unimaginable kinds may be brought closer. Conflicts among many different uses of the seas will become more frequent. Exploitation, if it occurs at all, will be monopolized by a few powerful states which alone have the technology and the military means to defend their activities. And political conflicts of many kinds, between rich states or enterprises competing for sites, between rich countries and poor over rights in the area, between countries and regions over boundaries, will become ever more frequent.

The issue thus provides a unique challenge to the international community. It is a challenge above all for the richer countries. They, above all, will be the ones that decide the issue. They may require to accept a régime that does not perhaps provide the maximum opportunities for themselves and their nationals, that may not even provide the most rapid possible exploitation of the sea-bed's resources.[1] They are not called on to be unduly generous. One of the great merits of this particular form of redistribution is that nobody needs to *give* anybody anything which they already possess. All they need to do is to refrain from *taking* what they do not have at present but might like to have in the future. Like the proposed creation of special drawing rights for the use of less developed countries within the IMF (another international measure that has so far been resisted by most richer countries) it is a way of *giving* something for nothing: a form of aid that would cost very little to the rich but mean very much to the poor. It is too, as in that case, one of the only forms of aid that has no direct balance of payments cost. The issue is thus a perfect test of whether the inhabitants of the richer parts of the world genuinely mean what they say when they talk of doing what they can to help poor countries. They could have no better opportunity of demonstrating this desire, at little direct cost to themselves, than they have here. Only a clear demonstration by their public opinions that something else is demanded of them is likely to induce governments to abandon their normal pursuit of individual national interests at all costs.

There are a number of specific issues on which the conflicts over the proposed régime may come to a head. But undoubtedly the most fundamental issue concerns the boundary between the national and the international areas. For whether the international régime can bring about any effective redistribution depends above all on where revenues for the régime begin to become

[1] But it is worth noting that it is the rich countries which will get most of the indirect benefits from sea-bed exploitation: new supplies of raw materials, more stable and lower prices, new opportunities and profits for their companies, new knowledge of the sea-bed and its resources, technological spin-off and so on, whatever type of régime is established.

available. There will be few substantial international revenues to be redistributed unless some royalties become payable in the areas richest in resources: that is, in those areas relatively close to the coast. The best way of achieving this would seem to be that suggested in this book, a way under which, outside the territorial sea (or even within it) up to 200 miles, royalties are *shared* between the coastal state and the international community, with the latter getting a progressively larger proportion as distance increases from the coast. This reflects a reasonable balance of interests and rights between coastal and non-coastal states.

This is the nature of the challenge, and the importance of the issues at stake. The governments and peoples of the richer countries especially have the opportunity here, if they choose, to rise above the pious platitudes they have so often uttered; to achieve in practice what they have so often professed in theory. They can cease desiring, hoping or promising to help. They can actually help—and in the most painless possible way.

But will they?

Index